Reviving Ancient Chinese Mathematics

Twentieth-century China has been caught between a desire to increase its wealth and power in line with other advanced nations, which, by implication, means copying their institutions, practices and values, whilst simultaneously seeking to preserve China's independence and historically formed identity. Over time, Chinese philosophers, writers, artists and politicians have all sought to reconcile these goals; this book shows how this search for a Chinese way penetrated even the most central, least contested area of modernity: science.

Reviving Ancient Chinese Mathematics is a study of the life of one of modern China's most admired scientific figures, the mathematician Wu Wen-Tsun. Negotiating the conflict between progress and tradition, he found a path that not only ensured his political and personal survival, but which also brought him renown as a mathematician of international status who claimed that he stood outside the dominant western tradition of mathematics. Wu Wen-Tsun's story highlights crucial developments and contradictions in twentieth-century China, the significance of which extends far beyond the field of mathematics. On one hand lies the appeal of radical scientific modernity, 'mechanization' in all its forms, and competitiveness within the international scientific community. On the other is an anxiety to preserve national traditions and make them part of the modernization project. Moreover, Wu's intellectual development also reflects the complex relationship between science and Maoist ideology, because his turn to history was powered by his internalization of certain aspects of Maoist ideology, including its utilitarian philosophy of science.

This book traces how Wu managed to combine political success and international scientific eminence, a story that has wider implications for a new century of increasing Chinese activity in the sciences. As such, it will be of great interest to students and scholars of Chinese history, the history of science and the history and philosophy of mathematics.

Jiri Hudecek is a Researcher at Charles University, Czech Republic.

Needham Research Institute Series
Series Editor: Christopher Cullen

Joseph Needham's 'Science and Civilisation' series began publication in the 1950s. At first it was seen as a piece of brilliant but isolated pioneering. However, at the beginning of the twenty-first century, it became clear that Needham's work had succeeded in creating a vibrant new intellectual field in the West. The books in this series cover topics that broadly relate to the practice of science, technology and medicine in East Asia, including China, Japan, Korea and Vietnam. The emphasis is on traditional forms of knowledge and practice, but without excluding modern studies that connect their topics with their historical and cultural context.

Celestial Lancets
A history and rationale of acupuncture and moxa
Lu Gwei-Djen and Joseph Needham
With a new introduction by Vivienne Lo

A Chinese Physician
Wang Ji and the Stone Mountain medical case histories
Joanna Grant

Chinese Mathematical Astrology
Reaching out to the stars
Ho Peng Yoke

Medieval Chinese Medicine
The Dunhuang medical manuscripts
Edited by Vivienne Lo and Christopher Cullen

Chinese Medicine in Early Communist China, 1945–1963
Medicine of revolution
Kim Taylor

Explorations in Daoism
Medicine and alchemy in literature
Ho Peng Yoke

Tibetan Medicine in the Contemporary World
Global politics of medical knowledge and practice
Laurent Pordié

The Evolution of Chinese Medicine
Northern Song dynasty, 960–1127
Asaf Goldschmidt

Speaking of Epidemics in Chinese Medicine
Disease and the geographic imagination in Late Imperial China
Marta E. Hanson

Reviving Ancient Chinese Mathematics
Mathematics, history and politics in the work of Wu Wen-Tsun
Jiri Hudecek

Reviving Ancient Chinese Mathematics

Mathematics, history and politics in the work of Wu Wen-Tsun

Jiri Hudecek

Routledge
Taylor & Francis Group

LONDON AND NEW YORK

First published 2014
by Routledge
2 Park Square, Milton Park, Abingdon, Oxon OX14 4RN

and by Routledge
711 Third Avenue, New York, NY 10017

First issued in paperback 2017

Routledge is an imprint of the Taylor & Francis Group, an informa business

British Library Cataloguing in Publication Data
A catalogue record for this book is available from the British Library

Library of Congress Cataloguing in Publication data
Hudecek, Jiri.
  Reviving ancient Chinese mathematics : mathematics, history, and politics in the work of Wu Wen-Tsun / Jiri Hudecek.
    pages   cm. – (Needham Research Institute series)
  Includes bibliographical references and index.
  1. Wu, Wen-tsün. 2. Mathematicians–China–Biography. 3. Mathematics, Chinese–History–20th century. I. Title.
  QA29.W82H83 2014
  510.92–dc23
  2014001682

ISBN 13: 978-1-138-09185-6 (pbk)
ISBN 13: 978-0-415-70296-6 (hbk)

Typeset in Times New Roman
by Out of House Publishing

Contents

Figures

Tables

Preface and acknowledgements

This is a book about mathematics in twentieth-century China. It is focused on a single mathematician, Wu Wen-Tsun (Wu Wenjun), in whose story we can see much of the overall development of twentieth-century Chinese mathematics. This story is more than a familiar tale of struggle, hard work and eventual achievement: it unites the individual career with the vicissitudes of politics and the power of tradition and history. Wu Wen-Tsun promotes the study of ancient Chinese mathematics and claims to revive its 'spirit' in his work – a claim which calls both for a critical scrutiny and contextual explanation. Why would Wu Wen-Tsun make use of ancient Chinese mathematics? Is it true that his method of mathematics mechanization uses ancient Chinese techniques? In what convoluted ways did twentieth-century ideologies enter mathematics and give respectability to historical enquiry even for an active mathematician? These are some of the questions that will run through the book.

This book is based on my dissertation 'You Fight Your Way, I Fight My Way: Wu Wen-Tsun and traditional Chinese mathematics', written at the Department of History and Philosophy of Science of the University of Cambridge and defended in February 2012. The PhD project was supported by the Needham Research Institute, Cambridge European Trust, and a Domestic Research Studentship of the University of Cambridge. Since 2012, I have been working on a project for the Ministry of Education of the Czech Republic, 'Deconstruction and construction of national traditions and science in China', which gave me the necessary peace of mind to revise the dissertation for publication. I would like to express my gratitude to all these institutions and their generosity.

The dissertation and book could not have been finished without the advice and help of many scholars and friends. First and foremost were my two joint PhD supervisors, Professor Catherine Jami of CNRS in Paris and Dr Eleanor Robson, who has since migrated from Cambridge to University College London. Their assistance has been essential from the very start, even before I actually embarked on the PhD project. The most successful parts of this book are products of their attention and criticism.

I owe no less gratitude to Professor Christopher Cullen. He supervised my earliest work on Wu Wen-Tsun, an MPhil dissertation in 2008, and in his capacity as Director of the Needham Research Institute (NRI) exerted huge effort to guide me and provide me with intellectual and material resources. He was especially crucial in persuading the publisher of the importance and interest of my book on this particular topic.

Among the most critical contributions of Catherine Jami to the writing of this book was to introduce me to Professor Li Wenlin of the Academy of Mathematics and System Science (AMSS) in Beijing. Professor Li, a historian of modern Chinese mathematics, sponsored my three stays at AMSS in 2008, 2010 and 2011, and arranged interviews with Wu Wen-Tsun as well as other senior mathematicians. Apart from his all-round generous hospitality, he also eagerly discussed my research with me and gave me many valuable hints and comments on the text.

I would also like to thank specifically John Moffett and Susan Bennett from the Needham Research Institute for their help throughout my stay in Cambridge, and the warm and friendly atmosphere they created in the NRI. Many more people have contributed at some stage to the research behind this book, and I can only mention some of them selectively: Professor Sir Geoffrey Lloyd, Dr Leon Rocha, Dr Lena Zuchowski and Dr Jacob Rasmussen in Cambridge; Hannah Mack from Taylor & Francis; Professor Karine Chemla in Paris; Professor Susanne Weigelin-Schwiedrzik and Dr Silke Fengler in Vienna; Professor Olga Lomova in Prague, and Professor Pavel Pech in Plzen; the late Professor Xu Yibao and Professor Joe Dauben in New York; Professor Doron Zeilberger from Rutgers University; Michael Barany from Princeton University; Fu Liangyu from the University of Pittsburgh; Dr Ying Jia-Ming from Taipei Medical University; Professors Lin Qun, Lu Qikeng, Wan Zhexian, Wang Yuan, Hu Zuoxuan, Shi He, Li Banghe and Gao Xiaoshan from AMSS, and Professor Guo Shuchun, Guo Jinhai, Liu Dun, Sun Chengsheng, Tian Miao, Xiong Weimin, Yuan Xiangdong, Zhang Baichun, Dr Zhao Zhenjiang, Dr Kubo Teryuki and Song Yonggang from the Institute for the History of Natural Sciences, Chinese Academy of Sciences (CAS); the staff of the CAS Archives; Professor Mei Jianjun, Dr Jiang Xi and Dr Yang Ruidong from Beijing University of Science and Technology; Dr Wang Xianfen from the Hebei Normal University; Professor Yi Degang from the Inner Mongolia Normal University; Dr Shao Kan from Jishou University; Professor Han Jishao and Dr Zhang Lujun from Shandong University; Dr Chen Pu from the Harbin Institute of Technology; Professor Ji Zhigang and his students Guo Yuanyuan and Wang Minchao from Shanghai Jiaotong University; Professor Hu Cheng, his students Liu Qiong and Deng Xiaojiao, and Dr Hao Xinhong from Nanjing University; Professor Xia Rubing from Nanjing Agricultural University; Professor Shi Yunli, Hu Huakai, Lü Lingfeng, Dr Ding Zhuojun, Fu Banghong, and the late Li Yu from the University of Science and Technology of China in Hefei.

Last, but not least, I would like to thank my parents for their patience and support, and my wife for being with me when I was revising the manuscript for publication. Where there are mistakes and bad arguments despite the help of so many excellent people, the responsibility is entirely my own.

Abbreviations

Abbreviation	Full title (Chinese equivalent where applicable)
AD	Academic Department (*Xuebu* 学部)
AMSS	Academy of Mathematics and System Science, Chinese Academy of Sciences (*Zhogguo kexueyuan Shuxue yu xitong kexue yanjiuyuan* 中国科学院数学与系统科学研究院)
ARP	Associate Research Professor (*Fuyanjiuyuan* 副研究员)
C.R. Acad. Sci. Paris	Comptes rendus hebdomadaires des séances de l'Académie des sciences de Paris
CAS	Chinese Academy of Sciences (*Zhongguo kexueyuan* 中国科学院)
CMS	Chinese Mathematical Society (*Zhongguo shuxue hui* 中国数学会)
CNKI	China Digital Knowledge Infrastructure (*Zhongguo zhishi jichu sheshi gongcheng* 中国知识基础设施工程)
CNRS	Centre National de la Recherche Scientifique
CPC	Communist Party of China (*Zhongguo gongchandang* 中国共产党)
CR	Cultural Revolution (*Wenhua da geming* 文化大革命)
CSCPRC	Committee on Scholarly Communication with the People's Republic of China
GLF	Great Leap Forward (*Da yuejin* 大跃进)
GS	Google Scholar
IHÉS	Institute des Hautes Études Scientifiques
IMCAS	Institute of Mathematics, Chinese Academy of Sciences (*Zhongguo kexuyuan Shuxue yanjiusuo* 中国科学院数学研究所)
IMU	International Mathematical Union
IPCAS	Institute of Physics, Chinese Academy of Sciences (*Zhongguo kexueyuan Wuli yanjiusuo* 中国科学院物理研究所)
ISSCAS	Institute of Systems Science, Chinese Academy of Sciences (*Zhongguo kexueyuan Xitong kexue yanjiusuo* 中国科学院系统科学研究所)
KMT	Kuomintang / National Party (*Guomindang* 国民党)
MM	*Mathematics Mechanization*
OICP	Out–In Complementary Principle (*Chu ru xiang bu yuanli* 出入相补原理)

Abbreviation	Full title (Chinese equivalent where applicable)
PDE	Partial differential equations
PLA	People's Liberation Army (*Renmin jiefangjun* 人民解放军)
PRC	People's Republic of China (*Zhonghua renmin gongheguo* 中华人民共和国)
RP	Research Professor (*Yanjiuyuan* 研究员)
SCCPCC	Standing Committee of the Chinese Political Consultative Conference (*Zhongguo zhengzhi xieshang huiyi changwu weiyuanhui* 中国政治协商会议常务委员会)
SNUC	Shanghai National University of Communications (*Shanghai guoli jiaotong daxue* 上海国立交通大学)
USTC	University of Science and Technology of China (*Zhongguo kexue jishu daxue* 中国科学技术大学)
WoS	Web of Science

1 Introduction

The history of twentieth-century China is punctuated with outbreaks of resistance against wholesale adoption of Western models of modernity. Some reached the form of organized political campaigns, culminating in certain phases of the Cultural Revolution (1966–76). But most remained on the level of conceptual alternatives formulated by individual thinkers and debated by public intellectuals. These repeated debates reflect the problematic compatibility of two goals simultaneously pursued by Chinese nationalism. One is to increase China's wealth and power to the level of the most advanced nations, which implies copying their institutions, practices and values. The other is to preserve China's independence and historically formed identity, without which nationalism loses any meaning.

What kind of synthesis could reconcile these two goals? Chinese philosophers, writers, artists and politicians all tried to come up with answers, which have been extensively studied.[1] But it has only recently been noticed how the search for a Chinese way penetrated even the most central, least contested area of modernity – science, an institution widely perceived in China as both completely foreign and completely positive.[2]

This book tells the story of an alternative Chinese mathematics, constructed or 'revived' not by a marginal mystic, but by one of China's most productive and admired mathematicians, Wu Wen-Tsun (Wu Wenjun 吴文俊 – see the note on transcription at the end of this chapter). This revived Chinese mathematics also successfully rejuvenated Wu's own career, paralyzed by the political turmoil and long isolation from international developments in post-1949 People's Republic of China.

Wu Wen-Tsun (born 1919) entered Chinese mathematics shortly after the Anti-Japanese War (1937–45), just as mathematics had become firmly established in China as a vigorous research field. Wu drew upon the determined vision and international contacts of his first teacher Chern Shiing-Shen (1911–2004), and quickly demonstrated his talent in algebraic topology, a dynamic field after World War II.

Wu Wen-Tsun spent four years in France in the leading centres of his discipline and returned to China in 1951 as a coveted prize for the new Communist regime, which was trying to enlist patriotic scientists working abroad to help with the construction of a new China. But the disruption of Chinese science in the Great Leap Forward (1958–60) and especially the Cultural Revolution (1966–76) cut off Wu from his peers abroad and effectively ended his topological career. By the time he could resume research in the mid-1970s, a huge gap had built up between his knowledge and the state of the art in algebraic topology. So Wu started a new twin research programme instead – 'mechanization of theorem-proving' and, at the same time, a study of traditional Chinese mathematics. These efforts soon led to the Wu method of mechanization, which became recognized as a breakthrough by the international artificial intelligence community.

Wu Wen-Tsun situated his success within a nationalist framework of independent modernization, and at the turn of the millennium became a government-promoted celebrity for this reason. Against the standard 'national hero' story told about Wu, this book portrays his turn to the history of Chinese mathematics as a sophisticated negotiation between the two conflicting goals of Chinese nationalism mentioned above, provoked by obstacles to mainstream academic mathematics in Mao Zedong's China.

Wu Wen-Tsun's reorientation to an independent research path, presented as a development of traditional Chinese mathematics, was an act of symbolic significance both for Chinese nativism and for the advocates of science and modernity. His life and career have been used to defend these positions, which makes it both challenging and worthwhile to critically re-examine the historical record. Wu Wen-Tsun's biographies published in China have generally avoided this task.[3] Before saying more about the approach of this book, let me sketch the contexts in which Wu Wen-Tsun has been discussed.

'You fight your way and I fight my way'

On 19 February 2001, the Chinese President Jiang Zemin awarded Wu Wen-Tsun the Highest National Science and Technology Award. The awards were established to promote Chinese 'independent innovation' (*zizhu chuangxin* 自主创新),[4] and the theme was also stressed in Wu Wen-Tsun's profile in the press release of the *New China* agency:

> In the late 1970s, against the background of great development of computer technology, he carried forward (*jicheng* 继承) and developed the tradition of Chinese ancient mathematics (algorithmic thought), and turned to the research of mechanical proofs of geometric theorems. This completely changed the discipline: it was a pioneering work within the international automatic reasoning community. It is called the 'Wu method', and has had great influence internationally.
>
> (Xinhua News Agency 2001a)

Wu Wen-Tsun claimed that his understanding of Chinese traditional mathematics enabled him to achieve a breakthrough in mechanical theorem-proving.

This was so, Wu believed, because he approached the problem of automated proofs from a different angle than standard modern mathematics, inspired by traditional Chinese mathematics. He was able to succeed where others before him had failed, because he did not blindly follow established models from abroad.

This attitude was in line with the famous Maoist slogan of seeking 'independence and self-reliance, regeneration through our own efforts' (*duli zizhu, zi li gengsheng* 独立自主，自力更生). This was retained by the Communist Party of China (CPC) as part of the 'living soul of Mao Zedong Thought' after the Cultural Revolution.[5]

But Wu more often chose another Maoist slogan to sum up his rejection of a 'universal' model:

> [During the Cultural Revolution] I read Mao Zedong's *Selected Works* from cover to cover. I started to side with Mao Zedong's view. You fight your way, I fight my way, the enemy advances, I retreat, the enemy retreats, I pursue. I think it is the same with mathematics. The enemy in mathematics is nature (*ziranjie* 自然界).
>
> (Wu Wen-Tsun, interview with the author, Beijing, 10 July 2010)

Wu described his scientific work in terms of warfare, since engagement in some kind of 'struggle' was the only legitimate activity in Maoist China. The words 'You fight your way, I fight my way' (*Ni da ni de, wo da wo de* 你打你的，我打我的) describe Mao's flexible combination of guerrilla tactics and regular warfare, which he employed in the many years of the Civil and Anti-Japanese Wars.[6] Mao's injunction also stresses strategic and innovative thinking, necessary in the backward conditions of developing China, as captured by Wu's allusion to the famous '16-word rhyme': 'The enemy advances, we retreat; the enemy camps, we harass; the enemy tires, we attack; the enemy retreats, we pursue.'[7] For Wu, Mao's definition of 'fighting my way' meant remaining the active party in conflict, always forcing 'my' terms on the enemy, withdrawing from engagements not to 'my' advantage.

The ability to remain active has been Wu Wen-Tsun's chief concern since his return to China in 1951. He knew that he remained dependent on developments in world mathematics, to which he had only limited and intermittent access. He thus sought a perspective that would enable him to 'fight with mathematical nature' independently, using locally available resources. To his delight, he found this perspective through his study of traditional Chinese mathematics.

Chinese intellectuals have often debated these questions: Did China produce anything similar to modern science? Did Chinese intellectual heritage lack essential components of modern scientific method, such as logic or willingness to experiment? On the other hand, did it also include unique features which had contributed to 'Western' science or would do so in future? The last point was often taken up by those disappointed with modern civilization and fearful of problems generated by purely 'Western' science.[8] Problems of this sort have often been subsumed under the so-called 'Needham question' (*Li Yuese nanti* 李约瑟难题), which became especially attractive for Chinese historians and philosophers of science after the end of the Cultural Revolution.[9]

The Needham question was the central concern of Joseph Needham's monumental project *Science and Civilisation in China*, and was initially formulated quite narrowly:

> How could it have been that the weakness of China in theory and geometrical systematization did not prevent the emergence of technological discoveries and inventions often far in advance (…) of contemporary Europe (…)? What were the inhibiting factors in Chinese civilization which prevented a rise of modern science in Asia analogous to that which took place in Europe from the 16th century onwards (…)?
>
> (Needham 1954: 3–4)

It is important to note that the question had a negative aspect (China's failure to generate modern science), but also a positive, affirmative part (China's successful application of natural knowledge throughout the pre-modern period), which made it attractive for various types of Chinese modernizing patriots (Amelung 2003: 251).

Many Chinese scholars have focused on the negative aspect of the Needham question, searching for the causes of the backwardness of Chinese science in the early modern period, or sometimes on the inner 'deficiencies' of traditional Chinese science and technology (Fan Dainian 1997), which implied the unsuitability of this traditional heritage for the development of modern science. This attitude prevailed even though the study of Chinese traditional knowledge has been supported by the state in the People's Republic of China since the early 1950s. The clearly stated goal of this effort was to increase national self-confidence and patriotism.[10] Although another goal was to assist economic construction by preserving useful traditional techniques and other knowledge,[11] these were clearly subordinated to modern science. Several famous scientists were drawn into this enterprise, such as the meteorologist Zhu Kezhen 竺可桢 (Chu Co-ching, 1890–1974), the physicists Ye Qisun 叶企孙 (1898–1977) and Qian Sanqiang 钱三强 (1913–92), and even Wu Wen-Tsun's boss at the Institute of Mathematics of the Chinese Academy of Sciences (IMCAS), Hua Loo-Keng 华罗庚 (1910–85).

The interest in Chinese traditional science in this period usually went hand in hand with an uncritical attitude towards modern science. In many cases, Chinese traditional science was taken as a vehicle for popular dissemination of modern scientific knowledge. It would be much more significant for patriotism, self-confidence and the credibility of nationalist models of development if someone could use core theoretical concepts or methodologies of Chinese traditional thought for new inventions or discoveries, rather than simply appropriate isolated techniques or results into the context of modern science.[12]

In recent years, some authors have tried to use Wu's success to demonstrate the viability of Chinese traditions in the modern world. Compared to these bolder arguments, simply presenting Wu as an accomplished model of independent innovation, as Wu Wen-Tsun's biographies written in China generally do, is a relatively modest and fact-based approach.

A good example of the recent trend is the book *The Rebirth of Oriental Science Culture* (Dongfang kexue wenhua de fuxing 东方科学文化的复兴; Zhu Qingshi and Jiang Yan 2004), to which Wu Wen-Tsun wrote a lengthy preface. The book contrasts 'Western scientific thought', based on reductionism and axiomatization, with the holism of 'Chinese traditional thought'. Reductionist thought is portrayed as futile in truly modern science. It cannot resolve the paradoxes of the theory of relativity and quantum theory, or overcome the logical limits imposed by Gödel's theorem (interpreted to mean that complete reduction to axioms is impossible). Reductionism and Western scientific thought in general are also made responsible for the multiple crises of Western society, echoing widespread post-World-War-I concerns which fuelled the 1923 Chinese debate about 'Science and the Philosophy of Life' (Chou 1978). Unlike the 1923 humanist opponents of scientism, however, the book suggests remedying the shortcomings of modern science by what it calls the 'Oriental scientific thought', which 'has been spontaneously converging with the development tendencies of modern science, and exhibits a high degree of consistency with many of its results', as 'many scholars are finding out' (Zhu Qingshi and Jiang Yan 2004: 215).[13] It is Wu Wen-Tsun's vindication of ancient Chinese knowledge and methodology in modern mathematics that serves as the only specific example in the book of a direct link between 'Oriental science' and modern scientific results.

Other authors place Wu Wen-Tsun into a culturally relativist framework, comparing the significance of his work to that of Thomas Kuhn's *Structure of Scientific Revolutions* (Hao Xinhong 2009b). If Kuhn introduced the notion of time-incommensurability, Wu added the dimension of space- or location-incommensurability. Not only is it impossible to judge the older theory by the standards of the newer one (and vice versa), but also conceptual systems from different civilizations should be evaluated purely on their own terms.[14] Wu's success is seen as a model to be followed by other scientific disciplines, in order to make full use of the specific character of Chinese science:

> Wu Wen-Tsun's mathematical breakthrough must encourage all knowledgeable Chinese, because the mechanization of mathematics has substantially absorbed the thought-style of our traditional mathematics and opened up a feasible, effective road to the renaissance of Chinese mathematics. It is a successful model of independent innovation in contemporary China, and he became a pioneer in opening the road of independent innovation for Chinese science.
>
> (Hao Xinhong 2009a: 11)

But Wu Wen-Tsun in fact remains unique, especially when compared to further examples of independent innovation on the basis of traditional Chinese knowledge (Li Shihui 2005).[15] Li found six cases of the 'complementarity of Western and Chinese cultures'. Wu Wen-Tsun is the first one. The second one is the seismologist Weng Wenbo 翁文波 (1912–94), who developed a predictive theory of earthquakes and other natural disasters on the basis of the Chinese calendrical cycles, such as the system of heavenly stems and earthly branches (*tian gan di zhi* 天干地支). Unlike Wu's method, however, Weng's theory has been rejected by fellow geophysicists in China, because 'his methods produce several solutions which have to be filtered out, otherwise the prediction accuracy decreases' (Li Shuhui 2005, quoting Ma Zongjin 马宗晋). [16] The third example is Ren Zhenqiu 任振球 (born 1934), a meteorologist who predicts natural disasters on the basis of astronomical phenomena apart from the annual solar cycle, using Chinese folk and literary sources. The fourth one is Chu Deying 褚德萤, who claims to have found biological changes in humans induced by meditations of *qigong* masters. The fifth case, Xu Qinqi 徐钦琦, explains the cycle of biological events in palaeontology by the waxing and waning of *yin* and *yang* in a 'big year'. The last representative of East–West fusion is Zhao Shaokui 赵少奎, a military engineer who has solved several problems in the design and testing of guided ballistic missiles by 'proceeding from China's concrete condition, "You fight your way, I fight my way"' and by adopting a holistic approach.

On this list, only Wu Wen-Tsun has created a theory both recognized by mainstream colleagues and linked to a sufficiently specific element of the Chinese tradition.[17] The significance of his story of inspiration from traditional Chinese mathematics for the creative power of Chinese traditions is obvious. On the other hand, Wu is also portrayed as an example of a 'scientific historian of mathematics', a proponent of sound methodology and impartial evaluation of Chinese mathematical heritage. He had been celebrated for this even before becoming famous amongst the wider Chinese public due to the award of 2001.

Wu Wen-Tsun as a historian of mathematics

Wu Wen-Tsun had an important influence on the community of professional and amateur historians of Chinese science, which was expanding rapidly in the 1980s (Xi Zezong 2000; Zhang Baichun 2001). The appraisals of Wu as a historian of Chinese mathematics have a different focus than the celebrations of his independent innovation: they show him as a pioneer of a sophisticated methodology.

Wu's colleague from the Institute of Mathematics, Li Wenlin 李文林 (born 1942), who switched from a research mathematics career to the history of mathematics in the same period that Wu Wen-Tsun became interested in traditional Chinese mathematics, pointed out in Li Wenlin (2001) the characteristics of Wu's approach to the history of Chinese mathematics. According to

Li, Wu first of all powerfully asserted the contribution of the special style of traditional Chinese mathematics to mainstream world mathematics. Wu also established a 'scientific method for studying the history of mathematics' when he demanded that only concepts and techniques indigenous to the tradition and time under study can be used in explanations. Wu formulated two basic principles for explaining historical mathematics: strict adherence to original texts and use of only those explanatory devices available to the authors of these texts. Wu was also a model of making 'the past serve the present' (*gu wei jin yong* 古为今用) as he absorbed inspiration not only from his studies of traditional Chinese mathematics, but also of Descartes' *Géometrie* and Hilbert's *Grundlagen der Geometrie* in his work.

'The past serving the present' is also a Maoist slogan, invoked in the 1970s in political contexts (see Chapters 4 and 5). But in Li's carefully structured argument, it is distinct from mere patriotic self-reliance, as Wu has taken the history of mathematics *in general*, not only of Chinese origin, for his inspiration. Wu's historical interest did indeed start with Western mathematics. At several moments of his life he engaged with older (though not ancient) Western mathematical literature and derived inspiration from it, as will be pointed out in later chapters.

The 'Wu paradigm' (Qu Anjing 2005) has been characterized in terms of the questions asked about traditional Chinese mathematics ('how was it done?' instead of the older 'what was done?'). But it certainly includes a utilitarian aspect too. Or perhaps, more precisely, the earlier, pre-Wu patriotic historiography instrumentalized history only for the promotion of national pride in ancient achievements, whereas Wu shifted attention to a more sophisticated practical utilitarianism, a search for ideas relevant to modern mathematical development. This programme has been carried forward by others who have made attempts to look at traditional Chinese mathematics as a whole and consider what is valuable about it for modern research (Wu Yubin 1987; Guo Jinbin 1987; Fu Hailun 2003).

Such approaches create a strong incentive for anachronism, but Wu himself combined his utilitarianism with a methodological purism, expressed in the often-quoted principles listed by Li Wenlin among Wu's achievements. According to Wu, any conclusions must be drawn from a first-hand analysis of extant historical texts, and modern tools and concepts must be left out of the interpretation. This may seem somewhat paradoxical. It is indeed difficult to reconcile the requirement for historical accuracy with a search for modern relevance. But Chinese historians of mathematics eager to claim Wu as an intellectual model never openly worry about this tension. It might be said that Wu has generated the interest he has by offering two very different perspectives, each attractive on its own: on the one hand, a present-centred focus and problem statement, and on the other hand, a 'scientific methodology' which should give the results more credibility.[18] This makes it especially interesting to study how these two principles were formed and combined in his own work.

Wu Wen-Tsun's sanguine appraisal of traditional Chinese mathematics' contributions to mainstream mathematical development reflected rather than formed scholarly opinion in China (see Chapter 5). But his position in the mathematical community allowed him to turn this into a powerful rhetorical gesture, intended to redress the imbalance created by Western historians' cavalier neglect of Chinese mathematics. Wu himself acknowledged in an interview with the author that his earliest claims about the contributions of Chinese mathematics, written in 1975, were deliberately exaggerated. But later, refined, versions of Wu's argument, focusing on the mechanized, algorithmic nature of Chinese mathematics as its specific strength, had an influence on the wider public in China, and also on younger historians. When cleared of their sinocentric bias, they even influenced Western historians of mathematics.[19]

In the slogan 'the past serving the present', the two words 'past' (*gu* 古) and 'present' (*jin* 今) form a dichotomy that is also reflected in Wu Wen-Tsun's label for the Chinese mathematical heritage. In his first articles,[20] he used the term *Zhongguo gudai shuxue* 中国古代数学, literally 'China's ancient mathematics'. But 'ancient' is not a very good equivalent of the Chinese *gudai*, which loosely indicates provenance from the past as opposed to the broadly understood present. Indeed, Wu's period of interest covered both the truly ancient texts,[21] and the sophisticated works of the thirteenth- and fourteenth-century mathematicians Qin Jiushao 秦九韶 and Zhu Shijie 朱世杰. Since about 1987, Wu has also used 'Chinese traditional mathematics' *Zhongguo* (or *wo guo*) *chuantong shuxue* 中国（我国）传统数学, starting with the article 'A new understanding of Chinese traditional mathematics' (Wu Wen-Tsun 1987b), or combinations such as 'traditional mathematics of ancient China' *Zhongguo gudai chuantong shuxue* 中国古代传统数学 (Preface to Wu Wen-Tsun 1996). On the other hand, Wu continued to use the adjective 'ancient' in his later book (Wu Wen-Tsun 2003) and in all interviews I am aware of – including an interview with me in July 2010.

In this book, I will usually talk about 'traditional Chinese mathematics', even though Wu used the term less frequently and might even have abandoned it completely after *c.*1996. One reason is that there is no good equivalent of *gudai* (unlike 'ancient', it can refer to any period before the modern), but more importantly, 'traditional Chinese mathematics' evokes the concepts of continuity and well-defined character, features which Wu sought more than anything else. Wu Wen-Tsun excluded from his definition of traditional or ancient Chinese mathematics the hybrid forms created after the arrival of Jesuits to China at the end of the sixteenth century. It is therefore better to call this mathematics 'traditional', as opposed to non-indigenous, Westernized mathematics, rather than the 'modern'.[22]

The central agument of this book is that Wu Wen-Tsun's interest in pre-modern Chinese mathematics was deeply imbedded in his appreciation of sustained tradition as a crucial component of the development of mathematics. This appreciation grew in the turbulent years of the late Maoist era, as a counterpoint to the discontinuities and disruptions which Chinese science had to endure in this period. But it also grew out of Wu's initial contact with the firmly established and continuous mathematical culture in Europe. Let me now sketch the structure and methodology of this argument as it will be presented in the rest of the book.

Wu Wen-Tsun as the subject of history writing

The focus of this book is Wu Wen-Tsun's reorientation in the late 1970s, the sources of that event, and the way it has been presented by Wu. Although the book is not intended to be a comprehensive biography, it is nevertheless arranged roughly chronologically, with a particular theme dominating each chapter.

Chapter 2 describes Wu Wen-Tsun's rise to mathematical prominence. Starting from his youth and early adulthood, it traces his studies with his first supervisor Chern Shiing-Shen (Chen Xingshen 陈省身), the stay in Strasbourg and Paris (1947–51) under the direction of eminent French mathematicians Charles Ehresmann and Henri Cartan, and the first successful years after his return to China (1951–8). It demonstrates the range of influences shaping the young Wu Wen-Tsun and his intimate experience of the international mathematical community.

Chapter 3 moves from a biographical account to the institutional and political setting of Wu's career in the late 1950s and early 1960s. I argue that the complicated developments of these years motivated Wu's later eagerness to find ways out of the mathematical mainstream, but also equipped him with some of the intellectual means to do that. Wu worked in the Institute of Mathematics of the Chinese Academy of Sciences (IMCAS, *Zhongguo kexueyuan shuxue yanjiusuo* 中国科学院数学研究所), under gradually increasing Party control and a series of political movements and policy changes. The Great Leap Forward inaugurated radical policies of 'linking theory with practice' (*lilun lianxi shiji* 理论联系实际), which remained permanently present even in the period of 'regularization' after the failure of the Great Leap Forward in 1960. Wu Wen-Tsun's response to these utilitarian requirements – studies of topological problems in game theory – are contrasted with approaches taken by some of his colleagues. The 1960s witnessed Wu's return to algebraic topology, which he also tried to promote at the newly established CAS-affiliated University of Science and Technology of China (USTC, *Zhongguo kexue jishu daxue* 中国科学技术大学). It was during this teaching that he contemplated questions of mathematical philosophy, such as the opposition of conceptual and computational mathematics, which he later tried to resolve by a turn to traditional Chinese mathematics.

Chapter 4 attempts to reconstruct, on the basis of painfully incomplete evidence, the history of IMCAS and Wu Wen-Tsun during the Cultural Revolution. This period falls into two parts, the chaotic and violent campaigns of 1966–71, and the more complex and better documented years prior to the launch of Deng Xiaoping's reforms in 1978. Given the incomplete record, we are left to speculate how the personal tragedies of his colleagues, material and ideological constraints on research, and special experiences such as factory work, formed Wu's decision to shift research course and, more significantly, enter the public sphere to legitimize this shift historically. An important factor was the opening of China to international scientific contacts in the 1970s, which demonstrated the increasing gap between current research and Chinese mathematical knowledge.

Chapter 5 is devoted to Wu Wen-Tsun's specific brand of historicism and analyses its intellectual sources. It shows that he was spurred to historical studies primarily by the work of Western historians of mathematics and their dismissive treatment of Chinese mathematics, although he also absorbed insights from Chinese scholars and Joseph Needham's work. The chapter stresses that Wu Wen-Tsun endowed his concept of 'traditional Chinese mathematics' with those characteristics that were also normatively positive within his own philosophy of mathematics.

Chapter 6 takes a close look at Wu Wen-Tsun's method of mechanization of mathematics. I analyse its establishment and propagation, compare its impact with the influence of Wu's earlier work on algebraic topology, and conclude that Wu's method is more widely known and used than Wu's formulae of the 1950s. The chapter also supports Wu's crucial claim that he had been inspired by traditional Chinese mathematics. The analysis of the structure of Wu's method identifies a key component which has a model in a traditional Chinese mathematical technique. I believe that this connection has not been pointed before with equal specificity. On the other hand, the chapter also assesses the contribution of Wu's acknowledged Western sources, i.e. the Ritt theory of algebraic differential equations, as crucial for the rigorous formulation of Wu's method. Chapter 6 concludes with a brief description of Wu's rising public status in China and the changed conditions of his work after 1979 in the newly established Institute of Systems Science of the Chinese Academy of Sciences (ISSCAS, *Zhongguo kexueyuan xitong kexue yanjiusuo* 中国科学院系统科学研究所).

The book concludes with a more general reflection on the problem of nationalism in science, and its manifestations in the work of Wu and other Chinese mathematicians. I use the theories of nationalism from political science to explain Wu Wen-Tsun's nationalist reaction to 'uneven development' (Ernest Gellner) and his 'routes to historicism' (Anthony D. Smith).

The book is written as a contextualized narrative drawing on a variety of published material – Wu's articles and books in Chinese as well as Western languages, his biographies, interviews conducted by me or other researchers (Huang Zubin and Wu Wen-Tsun 2004; Zhang Zhihui, *et al.* 2008), reminiscences from the *festschrift* published on the occasion of Wu's ninetieth birthday (Jiang Boju, *et al.* 2010), etc. Written primary sources include the following categories:

1   Wu Wen-Tsun's mathematical publications (articles and books).
2   Wu Wen-Tsun's writings on the history and philosophy of mathematics, mostly written since 1975. Many were collected in the volume Wu Wen-Tsun (1996).
3   Archival materials about IMCAS (1950–79) and ISSCAS (1979–*c.*1990) from the Archives of the Chinese Academy of Sciences (*Zhongguo kexueyuan dang'an guan* 中国科学院档案馆; repositories Z370 and Z373, respectively). Although I had full access to both repositories, there are other documents related to both institutes which have either not yet been handed over to the Archives of CAS, or are classified as archives of the Presidium (*yuanbu* 院部) and currently unavailable. These have not been consulted. On the other hand, I have consulted documents relating to Wu Wen-Tsun's brief stay at the Academia Sinica (*Zhongyang yanjiuyuan* 中央研究院) in 1946–7, deposited in the Second Historical Archives of China in Nanjing (*Zhongguo di er lishi dang'an guan* 中国第二历史档案馆).

Oral history interviews were used as a supplementary technique. I conducted one interview with Wu Wen-Tsun and several with other CAS academicians, his somewhat younger coworkers Wan Zhexian 万哲先 (born 1929), Lu Qikeng 陆启铿 (born 1928), Wang Yuan 王元 (born 1930) and Lin Qun 林群 (born 1936). Wu's students and followers Li Banghe 李邦和 (born 1942), Shi He 石赫 (born 1942) and Gao Xiaoshan 高小山 (born 1960) also shared valuable memories and insights with me. Only part of the material thus collected could be incorporated; in many cases, the information was available in published sources, which are used as references instead.

The available sources are insufficient to enable a complete break from the self-promotional biographies in which most information about Wu Wen-Tsun has been presented. This poses some serious methodological problems. Being based on Wu's contemporary reminiscences, his post-2000 biographies follow Wu's own focus on certain episodes and aspects of his life, to the detriment of others. The interpretation of past events in these books is heavily coloured by Wu's perspective at the time of the interview, exaggerating the continuity of his lifelong interests and attitudes. Looking back on his life from the threshold of the twenty-first century, Wu Wen-Tsun repeatedly stressed what he considered the most valuable part of his career – the understanding of the qualities and special character of ancient Chinese mathematics. This vindicated his choice to stay in China, and gave meaning to his entire life in the country:

> I would not have been able to reach an understanding of ancient Chinese mathematics without the events of my life here in China. This gives me a lot of satisfaction. It means that I did the right thing, even though not consciously, but unknowingly I took the right path.
>
> (Ke Linjuan 2009: 161)

My narrative tries to break away from this teleological perspective, and consequently highlights the contradictory moments and neglected details.

Throughout the book, I try to evaluate the importance of Wu's mathematical work by a combination of published peer reviews and a quantitative citation-tracking approach. From the former sources, we can see who read Wu's papers and what developments those papers inspired. Wu's papers have been mostly reviewed in the journal *Mathematical Reviews*, available online through the network MathSciNet. I also draw on the *History of Algebraic and Differential Topology* (Dieudonné 1989), written by Jean Dieudonné, one of the key figures of twentieth-century French mathematics and intimately acquainted with the research programme to which Wu also contributed.[23] Dieudonné's book, written as a lengthy literature review rather than a real-world history, is invaluable for understanding the perception of Wu's works by his French colleagues.

The quantitative approach to Wu's publication record uses the recorded citations of his papers in several electronic databases. The first database is the abovementioned MathSciNet (http://www.ams.org/mathscinet/), an outgrowth of the journal *Mathematical Reviews*. For every individual publication, MathSciNet includes its review from *Mathematical Reviews* and links to reviews of other items in MathSciNet that refer to this publication. More recent items (published since 1997) also include hyperlinked reference lists, covering all citations to other MathSciNet-covered publications.[24] A direct comparison of the citation counts of older and more recent works is problematic. On the other hand, MathSciNet has reliable matching and its user interface facilitates filtering of irrelevant or unwanted citations (e.g. those by the same author or in reviews of similar works that do no actually cite the publication).

There is another mathematical database run by the European Mathematical Society, built in a way similar to MathSciNet around the journal *Zentralblatt-MATH*. It is, however, much smaller and only shows citations in reviews, which are, moreover, less frequent than in MathSciNet.[25] All Wu Wen-Tsun's works combined had only 21 citations in this database, so I did not include these results in this survey.

I have also searched two science (i.e. not mathematics) databases, the ISI Web of Science (WoS) and its competitor Scopus (run by Elsevier). These databases are weak in their coverage of mathematical literature, but strong in physics and other science and technology disciplines. Their results are thus a good measure of the impact of a particular work beyond mathematics. Most citations come from the Web of Science; Scopus only had citations for one article which was not covered by the Web of Science. A serious limitation of the Web of Science is that it only lists authors' initials. Due to the similarity of Latin transcriptions of Chinese names, specific papers must be searched one by one by title, rather than as a list of all papers by WT Wu or WJ Wu, for example. Neither WoS nor Scopus tracks citations of books.

By far the highest citation counts were obtained through Google Scholar (GS) searches. Since its launch in 2004, GS has expanded in size and coverage, and improved its algorithms to deliver more reliable results. However, it still often includes duplicate citations and slightly distorts the counts by links to preprints and unpublished material. It also has a less than ideal user interface.

Some research suggests that Google Scholar is relatively more relevant for mathematics and computer science than for other academic fields. Moreover, it is considered more comprehensive for citations in non-English languages than, for example, WoS (Harzing and Wal 2008).

The last database used was China Academic Journal Database, part of the China National Knowledge Infrastructure (CNKI) project. It includes full texts of more than 31 million articles published in Chinese journals, including more than 5 million from the period 1915–93. CNKI recently introduced cross-referencing functionality, showing all citations, references, and co-citations for all papers within the database. The citations from Chinese conference proceedings, doctoral dissertations and 'excellent master dissertations' are also included. Citations were found only for Wu Wen-Tsun's articles written in Chinese; the citations in conference proceedings and doctoral dissertations were also counted, whereas those in master dissertations were omitted from my survey.

In general, Wu Wen-Tsun's own citations of his papers have been deducted from the citation counts; this was, however, not possible for the WoS figures, which were only available as aggregate numbers of citations rather than links to the papers.

How to reconcile data from the different databases? It is not possible to simply add them together, as each of the databases has its own scale (the largest number of citations in MathSciNet, for example, is less than 7 per cent of the largest number of citations in Google Scholar). In principle, one has to do what historians always do in the face of conflicting evidence: factor in the known limitations and biases and reach some kind of consensus. At the same time, each voice tells its own story worth noticing.

The general bias of all databases, and thus of quantitative assessment as a whole, is towards the present. Recent citations are more likely to be recorded, and recent papers are more likely to attract recent citations.

Apart from the first director of IMCAS, Hua Loo-Keng,[26] no twentieth-century Chinese mathematician has yet received sustained scholarly attention. Factual information about many twentieth-century Chinese mathematicians is available in the series *Biographies of Modern Chinese Mathematicians* (*Zhongguo xiandai shuxuejia zhuan* 中国现代数学家传), edited by Cheng Minde and published in five volumes between 1994 and 2002. These books are very useful, but lack a critical and analytic perspective, especially as most entries for younger mathematicians were written by their students or are even autobiographical. Oral history of Chinese mathematics, which has gained popularity recently,[27] also for the most part expands the range of primary sources, but does not fill the gap in historical analysis. Comprehensive histories of modern Chinese mathematics have been written by Zhang Dianzhou

(1999) and Liu Qiuhua (2010), but their character does not give the space to discuss particular mathematicians and their motivations in much detail. This book is therefore one of the first attempts to describe in depth the situation of a modern Chinese mathematician from an outside perspective, and to go beyond uncritically accepted individual memories.

A note on transcriptions and Chinese characters

This book generally uses the standard Chinese Romanization (*pinyin*). Proper names, however, are spelled in earlier transcriptions if such forma are more widely used and recognized. Whenever such a non-standard transcription appears for the first time, it will be accompanied with the corresponding *pinyin*. Thus 'Wu Wen-Tsun' is Wu Wenjun 吴文俊, 'Chern Shiing-Shen' Chen Xingshen 陈省身, and 'Hua Loo-Keng' stands for Hua Luogeng 华罗庚.

This book covers a period when the People's Republic of China gradually implemented a reform of the Chinese script. The sources on which I draw are very inconsistent in their use of characters, ranging from full traditional to second-wave simplifications (introduced after the Cultural Revolution and abandoned in the 1980s). To avoid confusion, I use current standard versions of simplified Chinese characters throughout, even for pre-modern book titles and personal names. The only exceptions are titles of publications and personal names from outside the PRC, which are not simplified.

Notes

1 Classic treatments of early twentieth-century Chinese resistance to modernity include Levenson (1968), Furth (1976), Alitto [1979] (1986), Chang (1987), among many others. Whereas these publications often portray the search for alternatives as a sign of failure to modernize effectively, the recent trend is to see in the effort to create Chinese modernity an important act of cultural creativity, linked to global unease with certain aspects of the modern world – see, e.g., Liu (1995), Daruvala (2000), Zarrow (2006) or Fung (2010).

2 Views challenging the omnipotence of science were decisively defeated in the 1920s – see Kwok (1965) and Chou (1978). Recent attempts at restrained methodological critique of scientism, inspired by modern Western science and technology studies, e.g. Jiang Xiaoyuan and Liu Bing (2007), have met with only limited success in wider public.

3 The earliest biography was written by Wu's student Li Banghe on the occasion of Wu's seventieth birthday (Li Banghe 1989). The entry in the biographical dictionary (Cheng Minde 1994: 377–400), written by Hu Zuoxuan, stresses Wu's position in the history of modern mathematics and his contacts with famous mathematicians. The most extended biography to date is Hu Zuoxuan and Shi He (2002), written after Wu received the China National Prize for Science and Technology in 2001. The most recent biography (Ke Linjuan 2009) combines excerpts from the previous sources with a transcript of a long television interview with Wu (Li Xiangdong and Zhang Tao 2006).

4 For a description of the selection process for this award, see Qin Jie and Wang Li (2001). The political significance of the award is explained in the Premier's speech (Zhu Rongji 2001) and in the editorial (Xinhua News Agency 2001b).

5 In the *Resolution on Certain Questions in the History of our Party since the Founding of the People's Republic of China* (June 1981), the 'living soul of Mao Zedong Thought' is defined as 'seeking truth from facts' (*shi shi qiu shi* 实事求是), 'the mass line' (*qunzhong luxian* 群众路线) and 'independence and self-reliance' (*duli zizhu* 独立自主). An alternative English translation of *duli zizhu* is 'Independence and initiative', as in the title of Mao Zedong [1938] (1960), usually translated as 'The question of independence and initiative within the United Front' (Mao Tse-Tung 1965).

6 Useful introductions and scholarly assessments of Mao Zedong Thought (*Mao Zedong sixiang* 毛泽东思想) are Schram (1989) and Knight (2007). An official Communist Party of China interpretation is in Liu Haifan and Wan Fuyi (2006). There is also a lot of literature on Mao's strategic doctrine. He Taiyou and Zhang Zhongliang (1993) and Johnston (1996) are particularly useful, but it is perhaps easiest to start with Mao's works *On Guerrilla Warfare* (1937) and *On Protracted War* (1938), available in English translations. These primary and secondary sources do not mention or analyse the slogan 'You fight your way, I fight my way', but they give a clear enough picture of what it meant. It is somewhat unclear how this slogan entered common use, because it does not appear in the officially approved four volumes of Mao's *Selected Works*. It can be found, however, in larger collections of Mao's statements, e.g. Mao Zedong (1993–9, 2010). Mao used it as a command to his generals (Mao Zedong [1947] 1996, [1950] 1999), as well as a recommendation to a delegation of the Palestinian Liberation Organization (unofficial English translation: Mao Tse-Tung [1965] 1994). Mao usually added that this was the most important and only universally valid principle of warfare.

7 *Di jin wo tui, di zhu wo rao, di pi wo da, di tui wo zhui* 敌进我退，敌驻我扰，敌疲我打，敌退我追, designed by Mao and Zhu De in 1930 (Mao Zedong [1930] 1960); cf. Short (2004: 222).

8 Probably the most famous example was the 1923 controversy between cultural conservatism, represented by Zhang Junmai and Liang Qichao, and pragmatic scientism, heralded by Ding Wenjiang and Hu Shi. This debate on 'Science and the Philosophy of Life' (*Kexue yu rensheng guan* 科学与人生观), often called the 'Science and Metaphysics Controversy' (*Ke xuan lunzhan* 科玄论战), decisively won by the scientistic camp, was initiated by the perceived failure of Western scientific civilization to avoid the carnage of World War I, and also by the worries of prominent Western intellectuals about the loss of spiritual values in the modern world. The tendency to use Western analysis of the weaknesses of predominant modern ways of thought as supporting arguments for preservation and restoration of Chinese traditions was then repeated several times in various campaigns and debates in the twentieth century.

9 From the vast amount of literature on the Needham question – Wang Qian Guozhong (2000) lists almost 300 articles and monographs published in China about it just between 1980 and 1998 – the most useful, relatively recent review article is probably Liu Dun (2000). In Chinese, the collection of Liu Dun and Wang Yangzong (2002) brings together the most important sources about the origins of the Needham question, and some of the most influential answers and reactions to it.

10 The relation between patriotism and historiography of science in China has been explicitly asserted in Chinese sources on many occasions, but there is little secondary literature about this phenomenon. An incisive summary is Zhang Baichun (2001); see also Amelung (2003).

11 See, for example, the *Twelve Year Plan for Research in the History of Chinese Natural Sciences and Technology* from 1956, as reproduced in Xi Zezong and Guo Jinhai (2011: 119–24).

12 Chinese civilization had 'reached an advanced level' in four areas of knowledge in the pre-modern period: mathematics, astronomy, medicine and agriculture (Liao

Yuqun 2006: 3). All of these traditions have been searched for techniques and data in the twentieth century. The complex case of traditional Chinese medicine has been studied most extensively (Scheid 2002; Taylor 2005), and has some clear success stories, e.g. the famous discovery of the antimalarial drug artemisin in Chinese material medica (Hsu 2006). Ancient Chinese astronomy had its celebrated contribution to modern science through the meticulous record of supernovae, compiled by Xi Zezong 席泽宗 (1927–2008) and Bo Shuren 薄树人 (1934–97). One of their catalogues (Xi Ze-zong and Po Shu-jen 1966) was translated into English and published in *Science*.

13 The authors evidently meant to say that modern science spontaneously converges with Oriental thought, but their formulation makes sense as well, as 'many scholars' often only discover the Oriental thought after their achievements in modern science. A Western scientist sympathetic to Chinese traditions, often mentioned in Chinese sources, is the Swedish physicist and Nobel Prize laureate Hannes Alfvén (1908–95), who is believed to have said in 1988 that mankind had to return to Confucius' wisdom (Li Shihui 2005: 12).

14 Hao Xinhong's view has most recently moved to a slightly different position, where she sees Wu as striking a compromise between multi-culturalism and universal standards of validity, avoiding 'anything goes' relativism – see Hao Xinhong (2011).

15 I thank Hao Xinhong for pointing out this article to me.

16 Note, however, more recent articles by Chinese scientists still promoting the citation of Weng Wenbo, e.g. Ouyang Shoucheng and Peng Taoyong (2005).

17 There are many other attempts to build new 'scientific theories', especially on traditional Chinese divination techniques, with even less impressive credentials than those on Li Shuihui's list. Many verge on the crackpot variety (Jia Hepeng and Li Jiao 2007).

18 Wu's principles are sometimes directly quoted as a methodological guide (Wu Maogui and Xu Kang 1987). Wu is also cited as an example of a modern mathematician who sees the value of a non-presentist interpretation of ancient texts (Cullen 2004: 21–2).

19 Wu's belief that traditional Chinese mathematics represents an algorithmic trend in the general evolution of mathematics has been taken up by Li Wenlin (1991), and, to some extent, Karine Chemla (1987). Comparisons of ancient Chinese and ancient Greek mathematics, i.e. of the *Nine Chapters* and Euclid's *Elements*, are especially widespread, and often have a direct or indirect connection to Wu Wen-Tsun's writings.

20 At least 14 journal articles, speeches and prefaces starting with the pseudonymous Gu Jinyong (1975), and including the English-language talk at ICM 1986 (Wu Wen-Tsun 1986b).

21 The *Zhou bi* 周髀, sometimes rendered as *The Mathematical Classic of the Gnomon of Zhou*, probably from the first century BCE (see Cullen 1996), and the *Nine Chapters on the Mathematical Art* (*Jiu zhang suan shu* 九章算术), hereafter abbreviated as the *Nine Chapters*, probably from the first century CE (see Shen Kangshen, *et al.* 1999; Chemla and Guo Shuchun 2004).

22 The community of Chinese historians of mathematics is divided in their choice of terms. *Zhongguo chuantong shuxue* is preferred, e.g., by Li Di (1984); Guo Shuchun (2004); Fu Hailun (2003); Li Zhaohua in Zhu Shijie [1303] (2007); Guo Jinbin and Kong Guoping (2004). On the other hand, Li Yan and Du Shiran (1963–4), Qian Baocong (1964), Liu Dun (1993) and Li Wenlin (2005) use *Zhongguo gudai shuxue*. In Western scholarly literature, 'traditional' seems to be used more often when talking about the entirety of Chinese mathematics before the impact of the Jesuits (e.g. Martzloff 1997), whereas 'ancient' is used only for the specific period of the formation of the Chinese mathematical canon, as in Lam Lay Yong and Ang

Tian Se (2004) or (Dauben 2007). 'Traditional Chinese mathematics' also points to another nebulous, emotionally charged category, Traditional Chinese Medicine (TCM). As TCM, in China generally called simply 'Chinese medicine' (*zhong yi* 中医), became a cherished part of Chinese national identity, mathematics could draw upon its prestige by receiving a similar name.

23 Dieudonné was also the leading writer behind 'Nicolas Bourbaki', a group that had a major influence on Wu; see Chapter 5.

24 On MathSciNet, see http://www.ams.org/mathscinet/help/about.html, http://www.ams.org/mathscinet/help/citation_database_understanding.html (accessed 7 March 2011).

25 On ZentralBlatt-MATH, see http://www.zentralblatt-math.org/zbmath/about/ (accessed 7 March 2011).

26 See his obituary (Halberstam 1986), but especially the biography written by his student Wang Yuan in 1987, which has also been published in English (Wang Yuan 1999).

27 The best example is the carefully edited book by Xu Lizhi, *et al.* (2009). A book based on interviews with Hong Sheng, one of Hua Loo-Keng's students and later Vice-President of USTC, is currently under preparation by Zhang Zhihui. Some shorter interviews have been published in journals, for example with Duan Xuefu, head of the Department of Mathematics at Peking University – Ding Shisun, *et al.* (1994). Literature about Wu Wen-Tsun in fact also mostly belongs to the oral history category, based on his reminiscences, supplemented to a varying degree by additional explanation and commentary.

2 The making of a prominent Chinese mathematician

Early life (1919–45)

Wu Wen-Tsun was born in Shanghai on 12 May 1919 as the first child of four. His father Wu Futong 吴福同 was an educated professional in a medical publishing house. Wu Futong was a graduate of a new Western-style high school affiliated to Nanyang College (*Nanyang gongxue* 南洋公学), established in 1896 to train competent technicians necessary for China's modernization. Wu Futong was valued for his English language skills, and he and his family led a middle-class lifestyle. He had a large library and provided education for all of his children; however, only the eldest son Wu Wen-Tsun proceeded beyond lower-middle school. Wen-Tsun's younger brother Wu Wenjie 吴文傑 died in childhood, and their two sisters Wu Wenjuan 吴文娟 and Wu Wenmei 吴文美 never attended high school (Ke Linjuan 2009: 1–3).

Wu's younger brother died of the long-term effects of a head injury he had suffered while playing away from home. This made Wu Futong and his wife Shen Cuihua 沈粹华 particularly concerned about Wu Wen-Tsun's health, and prompted them to limit his activities out of family oversight. He thus spent most of his childhood in his father's library, reading all kinds of literature, especially both Chinese classical and translated Western novels, and historical works. He was influenced very strongly by classical Chinese satires of bureaucracy, which made him dislike official positions and bureaucratic apparatus (Hu Zuoxuan and Shi He 2002: 3–4).

Wu's primary and lower-middle school studies did not reveal any mathematical inclinations. They focused on languages, including classical Chinese, English and German. They were also interrupted by illnesses and – during lower-middle school – the Japanese military incursion into Shanghai in 1932. Wu spent several months of that year away from Shanghai (and his school) in a rural safe haven, and had trouble understanding mathematical classes when he came back. He gave up mathematics altogether and received 0 points on the final-year exam. His parents therefore sent him to supplementary summer classes, and it was there that he first felt interest in mathematics and learned the techniques of geometrical proof. Wu considered the teacher of this summer course his first guide to mathematics (Ke Linjuan 2009: 4–11; Zhang Zhihui, *et al.* 2008: 96).

Wu's mathematical interests started to develop more rapidly during his three years at high school, primarily thanks to a teacher of geometry, who assigned him a lot of difficult proof exercises. But Wu still had a much keener interest in languages (he was already reading unabridged English novels) and also in physics, especially mechanics. This was related to his good grasp of geometry, which underlies most high-school mechanics. On the other hand, he was not greatly attracted to the experimental aspect of physics. After finishing high school, Wu was recommended for a university scholarship for the Shanghai Section of the National University of Communications (*Guoli jiaotong daxue shanghai benbu* 国立交通大学上海本部, SNUC), but his eagerness for mechanics was ignored and the scholarship required him to study mathematics (Ke Linjuan 2009: 11–19).

The National University of Communications was a direct descendant of the Nanyang College, where Wu Wen-Tsun's father had studied a quarter of a century earlier. It later developed into a major supplier of technical personnel for government-run services, such as the postal and telegraph service and railroads. Being primarily a polytechnic, SNUC had founded a department of mathematics only in 1928 as a service department for the technical subjects. The first mathematics majors were admitted only in 1930. The subjects of the full four-year course are summarized in Table 2.1.

It is remarkable how much attention was paid to foreign languages and other humanities (including the Kuomintang (KMT) political education, which nevertheless occupied minimal space compared to politics in post-1949 curricula). Even more time was devoted to chemistry and physics, which alone accounted for 20 per cent of all credits and was taught with the same requirements as for the SNUC physics major. The list of mathematical lectures is, on the other hand, relatively limited, with no mention of statistics, for example.[1] Emphasis was placed on calculations, both in analysis and 'modern algebra', which in fact mainly involved matrix calculus. Wu Wen-Tsun found these classical mathematical subjects dull and wanted to abandon the discipline after his second year (Ke Linjuan 2009: 20–3).

In late 1937, Shanghai was captured by Japanese armies. A minor part of SNUC had moved in advance to inland China, but most teachers and students, including Wu Wen-Tsun, stayed in Shanghai and re-established the university in the French concession, an island of relative tranquillity until 1941, as long as Japan was only at war with China.

Wu Wen-Tsun lived in strained economic conditions after the move to the French concession, because the scholarship paid by his former high school was stopped. Wu demonstrated his patriotic idealism at this junction: The former principal of his high school Chen Qun 陈群 offered him a personal bursary, but Wu refused to accept it because Chen had become minister of the interior in the puppet government controlled by the Japanese (Ke Linjuan 2009: 20–6).

Despite these difficulties, Wu finally became truly interested in mathematics. The crucial trigger was the third-year course on functions of real variables,

Table 2.1 Subjects of the mathematics major at SNUC, 1936–40

Course name *(hours per week)*			
First year	*Second year*	*Third year*	*Fourth year*
Chinese (3)			
English (5)	German (5)	German (4)	
[KMT] Party Line 党义 (1)	History of scientific thought (3)	Economics (3)	
Physics (4)	Physics (4)	Theoretical physics (3)	Theoretical physics (3)
Physics – labs (3)	Physics – labs (3)	Electromagnetism (3)	Modern physics (3–4)
Chemistry (4)	Theoretical mechanics (4)		
Chemistry – labs (3)			
Calculus (4)	Advanced calculus (4)	Function theory (3)	Modern analysis A (3)
Equations (2)	Differential equations (2)	Real function theory (3)	Modern analysis B (3)
		Theory of infinite series (3)	Number theory/ Group theory (3)
		Modern algebra (3)	
Geometrical drawing/ Engineering drawing (3)	Analytic geometry in space (3)	Modern geometry (3)	Differential geometry (3)
Military training (3)	Sports (2)		
	Mathematical problems (2)	Mathematical problems (3)	Mathematical problems (4)
			Lectures by specialists (2)
			Dissertation (5)

Source: Jiaotong University History Group (1986: 237–8).

which introduced him to modern mathematics, including topology. Many courses in the third year were taught by Wu Chonglin 武崇林 (1900–53), who noticed Wu Wen-Tsun's talent and interest, and provided him with additional books, often not readily available in China at the time (Zhang Zhihui, *et al.* 2008: 96).

Wu Wen-Tsun's only classmate at SNUC was Zhao Mengyang 赵孟养. They spent much time together, mostly at Zhao's home, where Wu observed Zhao's father and brother playing *weiqi* (go) and became an avid player too. Zhao Mengyang had a decisive role for Wu's career after the war, when he introduced him to several famous Chinese mathematicians,[2] including Wu's first supervisor Chern Shiing-Shen (Chen Xingshen 陈省身 or S.S. Chern, 1911–2004). Zhao later became a teacher of mathematics and translator of Soviet mathematical literature (Ke Linjuan 2009: 50–1).

Wu Wen-Tsun finished his studies with a dissertation on 'A proof of the Pascal theorem by the method of mechanics'. He had already developed the basic idea in his first year, and expanded it to a collection of proofs of theorems about 60 Pascalian configurations, all with the same method based on considerations from dynamics. This topic was to remain a recurring interest – he published a short booklet on the mechanical method in elementary geometry (Wu Wen-Tsun 1962), and returned to the study of Pascalian configurations in the late 1970s in connection with studying David Hilbert's *Foundations of Geometry* (see Chapter 6).

After finishing university, Wu started teaching at a secondary school. Although only five mathematics majors graduated from SNUC between 1938 and 1940, there were no places for a fresh mathematician outside the education sector. Wu's first school eventually closed and he had to teach at even more elementary level until 1945, supplementing his income with administrative and supervisory work (Zhang Zhihui, *et al.* 2008: 97). He had no time to do research, as he continued to live with his parents and had to go to bed early. Thanks to his teaching, however, Wu derived a confident familiarity with the nitty-gritty details of Euclidean elementary geometry and its synthetic style of proof, which was to become important in the development of the Wu method in the 1970s (Ke Linjuan 2009: 34).

Becoming a famous topologist (1945–51)

The first six years after the Anti-Japanese War were crucial in Wu Wen-Tsun's life: most of his later work in algebraic topology was directly or indirectly related to the unique knowledge and skills he acquired in this relatively short period. He also established personal contacts with mathematicians whose work continued to inspire him throughout his mathematical career.

The part of SNUC which had managed to escape from Japanese occupation to the provisional capital of the Republic of China, Chongqing, returned to Shanghai in 1945 as the 'Provisional University' (*Linshi daxue* 临时大学). Wu joined the Department of Mathematics as teaching assistant on Zhao Mengyang's recommendation (Hu Zuoxuan and Shi He 2002: 26–7).

But Wu was soon attracted to the Preparatory Office for the Institute of Mathematics (*Shuxue yanjiusuo choubeichu* 数学研究所筹备处) of Academia Sinica, which moved to Shanghai after the war. The Institute had been in preparation since 1942 in wartime Kunming, but all its members had full-time jobs at universities, and sometimes even abroad. The director Chiang Li-fu (Jiang Lifu 姜立夫)[3] left China for the USA in May 1946, and actual leadership passed into the hands of S.S. Chern, then a professor at Tsinghua University, who had become a leading expert on differential geometry during his studies in Hamburg and Paris in the mid-1930s. S.S. Chern had spent the years 1943–5 at the Institute for Advanced Studies in Princeton and arrived back in Shanghai in April 1946. He turned the Preparatory Office into a kind of graduate school and searched among recent graduates of China's best universities for talented young mathematicians (Chern Shiing-Shen 1988: 14). Six years after his graduation, Wu Wen-Tsun no longer belonged to this category, but he was persuaded by Zhao Mengyang to approach Chern directly and ask for a job at the institute, a request which Chern accepted (Hu Zuoxuan and Shi He 2002: 31–3; Ke Linjuan 2009: 49). To recommend himself, Wu presented a list of eight works on elementary geometry, including his university dissertation and papers about relations between Pascal configurations in Desargues geometries. He worked in the institute unofficially from August 1946, with a formal appointment on 16 September (Institute of Mathematics of Academia Sinica (Shanghai) 1947: 132–3).

Wu also submitted to Chern a newly written review paper investigating the logical relationships between concepts of point-set topology. Chern considered this topic useless and returned Wu's report with the verdict 'wrong direction'. He warned Wu that he would never produce any truly good mathematical work with such purely logical investigations, and that he had to focus on problems with real geometric content instead. The future was in advanced combinatorial topology, which was turning into a new branch called algebraic topology (Wu Wen-Tsun 2006: 462).

After Wu had spent a few weeks reading mathematical literature in the library of the institute, Chern persuaded him to 'repay the debts' to the authors and write something in turn. Wu responded with his first paper (Wu Wen-Tsun 1947), on essential symmetric products of topological spaces. Chern managed to get the paper published in France via Élie Cartan (1869–1951), whom he knew well from his stay in Paris in 1936–7. Even more important for Wu's further career was a proof of a theorem about the Stiefel–Whitney characteristic classes of sphere bundles (Wu Wen-Tsun 1948a), which appeared in *Annals of Mathematics*, one of the world's leading mathematical journals, together with an accompanying paper by Chern.

It is impossible to explain precisely the meaning of the term 'characteristic class' without a lengthy introduction to algebraic topology. But since it is going to come up repeatedly in the discussion of Wu Wen-Tsun's work, the reader might still like to have an approximate idea of what it means. Let us imagine a smooth connected surface (such as the surface of a sphere), and attempt to cover it with arrows in such a way that each point will have a unique arrow assigned to it, and the direction of these arrows will change continuously from point to point, without abrupt changes of direction. If the surface is a sphere, this attempt will necessarily fail and we will always end up with one or two points where the arrows abruptly reverse their orientation (a 'source' and a 'destination'). This result is characteristic for the combination of a two-dimensional sphere and a field of one-dimensional arrows. Although the actual points can be in various places depending on the vector field and the actual shape of the sphere, and may even collapse to a single point that is both the source and destination of all arrows, something fundamental about this singularity is topologically invariant – if the original sphere is deformed and stretched as if made of rubber, the singularity will fundamentally remain the same. This abstract fundamental property is the characteristic class, in this case the one-dimensional characteristic class of the two-dimensional sphere. It is natural to extend this concept to other smooth connected spaces (manifolds) and to higher dimensions.[4]

The way in which Wu became interested in the topic is itself worth recording, as it illustrates the travel of ideas in mid-twentieth-century mathematics. Chern asked Wu to read a paper by Samuel Eilenberg in the proceedings from the University of Michigan topology conference of 1940 (Wilder and Ayres 1941). Wu was instead attracted to the next paper in the book (Whitney 1941) and applied techniques he had learnt in the other seminal paper on characteristic classes (Stiefel 1935) to prove a theorem Whitney announced without proof (Wu Wen-Tsun 2006: 465).

Wu's ability to achieve such a result and get an article published within a year of taking up algebraic topology later greatly surprised his French colleagues (Wu Wen-Tsun 1989a). Some authors claim that even Whitney was so convinced by the article that he did not find it necessary to publish or even keep the unpublished manuscript of his own proof, and indeed it has never been published (Li Banghe 2010: 4). But Wu's contribution was not regarded as a complete clarification at the time. In his review of Wu Wen-Tsun (1948a), Whitney observed 'This proof of the duality theorem is considerably simpler than that of the reviewer (unpublished)', but the review also observed that the proof was in fact incorrect (Whitney 1949). Norman Steenrod in his book *The Topology of Fibre Bundles* also did not consider Wu's proof definitive:

> The proposition [i.e. Whitney's theorem] is somewhat ambiguous (…). W.T. Wu has proved the special case obtained by reducing everything mod 2. Reduction mod 2 eliminates the ambiguity. (…) A clarification and proof in full generality is needed.
>
> (Steenrod 1951: 198–9)

Wu's first article on characteristic classes continued to inspire his research. On the other hand, while it is clear that many algebraic topologists were aware of this result and referred to it, searches in citation databases return only three independent citations recorded in Google Scholar, and none at all in MathSciNet. This highlights the general problems with tracking citations of relatively old publications.

Chern valued Wu's work highly, which can be seen both from his later published reminiscences and from positive comments he wrote about his student in a staff evaluation table at the time. He awarded Wu the highest bonus among all young researchers (Institute of Mathematics of Academia Sinica (Shanghai) 1947: 258).

Before entering Chern's institute, Wu had taken an examination for Chinese government scholarships to France. This was also on Zhao Mengyang's advice, with additional persuasion by Wu's professor at the Provisional University, Zheng Taipu 郑太朴 (1901–49). The examination selected 40 students in different disciplines. When successful candidates were announced in 1947, Wu had the highest score among the four mathematicians selected. They were called to Nanjing in the summer to receive language and other preparatory training (Zhang Zhihui, *et al.* 2008), and Wu thus left Chern's institute after only one year. Chern accepted a position at the University of Chicago soon afterwards, and did not return to China until 1972.

Chern recommended Wu to Henri Cartan (1904–2008), son of Chern's mentor Élie Cartan, who was then working in Strasbourg, a city Chern recommended for its relative tranquillity compared to Paris (Li Xiangdong and Zhang Tao 2006). But by the time Wu arrived in Strasbourg, Cartan had already been appointed a professor at the École normale supérieure in Paris. Wu thus instead became a doctoral student of Charles Ehresmann (1905–79), another student of Henri Cartan's father Élie. Ehresmann was a very influential mathematician and author of several important topological concepts. Wu Wen-Tsun found his approach close to what he was familiar with, and with hindsight considered the change of advisor very beneficial (Ke Linjuan 2009: 54). Wu based his dissertation about characteristic classes on concepts introduced by Ehresmann (especially 'spherical fibre structures') and frequently published short notes about this topic in the *Comptes Rendus de l'Academie des sciences de Paris*. The results of one of these articles (Wu Wen-Tsun 1948b) shocked Heinz Hopf (1894–1971), mentor of the founder of characteristic classes Eduard Stiefel, so much that he travelled personally to Strasbourg to question Wu about his method. Impressed with Wu's reply, Hopf invited him for a short visit to the Federal Polytechnic (ETH) in Zurich. During this visit, Wu met the Chinese topologist Jiang Zehan (Kiang Tsai-Han 江泽涵, 1902–94),[5] who later became head of the Department of Mathematics at Peking University and invited Wu to teach there in 1950 (Ke Linjuan 2009: 54–8).

Before Wu was awarded the scholarship to France, he had only a passive, self-taught knowledge of written French. This was supplemented by three months of intensive training in Nanjing prior to departure from China. He could understand technical terms during seminars, but preferred talking to other mathematicians in English. His contacts with French colleagues were consequently rather limited. Ehresmann rarely saw his students, usually only when they needed help or had derived new results. On the other hand, Wu befriended René Thom (1923–2002), who had studied with Henri Cartan, but stayed in Strasbourg after his teacher's departure. Wu introduced Thom to characteristic classes, and in this way contributed to Thom's Fields Medal (1958) for the establishment of cobordism theory (Thom 1990: 200).

Wu defended his dissertation 'Sur les classes caractéristiques des structures fibrées sphériques' ('On characteristic classes of spherical fibre structures') in July 1949. It investigated the properties of different constructions of characteristic classes: the Stiefel–Whitney classes, the Pontrjagin classes (Pontrjagin 1942) and the Chern classes, defined by a construction similar to the Pontrjagin classes but for more general complex manifolds (Chern Shiing-Shen 1946). Wu reformulated all known results about characteristic classes in the recently developed language of singular cohomology, and proved relationships between the different types of characteristic classes. This highlighted the fundamental character of the Chern classes, which could be used to calculate all others. The names Stiefel–Whitney, Pontrjagin, Chern and Euler classes, and the corresponding symbols w, p, c and e, were first established in Wu's thesis.

Wu Wen-Tsun's thesis took a relatively long time to publish. When it finally appeared (Wu Wen-Tsun and Reeb 1952), its results had already been established in the research community via his separate publications in the *Comptes Rendus de l'Académie des sciences*, presentations at Henri Cartan's seminar in 1950, and as independent developments by other scholars. This was noted by the reviewer: 'It seems regrettable that this thesis took so long to appear in print, particularly in view of the progress which has taken place in the meantime (due to the author and others)' (Samelson 1953). Nevertheless, its influence has been lasting. It has 82 citations on Google Scholar, the latest from 2012. MathSciNet records 20 citations up to 2013. This influence is probably due mainly to the references in the textbook by Milnor and Stasheff (1974), which advised the readers to 'consult [Wu's thesis] for further information' on cell structures for Grassmann manifolds. This textbook is widely cited (MathSciNet 601, Google Scholar 2551 citations), and probably served as a main channel for raising awareness of Wu's work.

Wu's most important work in France, however, only took place after he moved to Paris in 1949, having obtained a stipend to work at the Centre national de la recherche scientifique (CNRS) in Paris under the direction of Henri Cartan. René Thom, who also came from Strasbourg and shared an office with Wu, introduced him to a new tool – a squaring operation on cohomology classes, or 'Steenrod squares',[6] which was especially well suited to the algebraic structure of Stiefel–Whitney characteristic classes. This inspired Wu's work which introduced his name into mathematical textbooks in the form of the 'Wu formula', 'Wu theorem' and 'Wu classes'.

Thom and Wu used Henri Cartan's axiomatization of Steenrod squares to derive a set of formulas for Stiefel–Whitney characteristic classes. Thom's results (Thom 1950) are in a sense more fundamental, but Wu's two formulas were useful for immediate calculations of characteristic classes and found numerous applications. In Wu Wen-Tsun (1950a), he defined new characteristic classes U^p implicitly by the equation

$$U^p \cup Y^{n-p} = Sq^p Y^{n-p} \tag{2.1}$$

that is, by a Steenrod square (Sq^p) and cup product ($\cup$) of an arbitrary cohomology class Y^{n-p}. These classes are linked by the 'Wu formula'

$$W^i = \sum_{p=0}^{i} Sq^{i-p} U^p \tag{2.2}$$

to the Stiefel–Whitney classes W^i, again by Steenrod squares.

The newly created classes U^p are now called the Wu classes.[7] The Wu formula was used by Thom (1952a) to prove that Stiefel–Whitney classes are indeed topological invariants (Dieudonné 1989: 537).

In another short note (Wu Wen-Tsun 1950b), presented at the same session of the French Academy of Sciences, Wu announced and demonstrated by induction another formula, which specified the relationship between all Stiefel–Whitney classes of any vector bundle:

$$Sq^r W^s = \sum_{t=0}^{r} \binom{t}{s-r+t-1} W^{r-t} W^{s+t} \qquad\qquad s \geq r > 0 \qquad\qquad (2.3)$$

The number of indexes makes the formula look complicated, but its use is actually quite straightforward in the usual binary (mod 2) cohomology, whose coefficients are either 1 or 0. It was shown to be a complete description of all relationships between Stiefel–Whitney classes by Dold (1956).

Steenrod squares continued to attract Wu's attention even after 1950. Before leaving France, he used Thom's axiomatization of Steenrod squares to prove their equivalence with the Smith calculus (Richardson and Smith 1938). He later revisited this topic (Wu Wen-Tsun 1957c) and proved the equivalence directly.

Wu communicated with other great mathematicians of Cartan's group, such as Jean-Pierre Serre (born 1926, Fields Medal 1954), but seems never to have became a member of the French mathematical community. He always preferred English when publishing outside France, and even stated he preferred German mathematical literature to French writings, which were 'too disorderly and moving at a frantic pace' (Ke Linjuan 2009: 66). In spring 1951, he accepted an invitation from Jiang Zehan, whom he had met in Zurich and Strasbourg, to return to China and teach at Peking University. Henri Cartan thought that Wu could build a circle of students in China, and supported his decision. Ehresmann, on the other hand, tried to persuade Wu that he would get 'big chunks of money' if he stayed in the West (Li Xiangdong and Zhang Tao 2006: 36:30–37:40). In the end, Wu left for China from Marseille in July 1951. Shortly afterwards, a letter from the Princeton Institute of Advanced Studies reached the CNRS, inviting Wu to take up a teaching fellowship at Princeton, but Wu was already on his way.

The results from this period of Wu's career are by far the most important of his pre-1977 work. Interestingly, the citation numbers of his two 1950 papers are higher than those of the articles by Cartan and Thom in the same issue of *Comptes Rendu*, which introduced the underlying theory (see Table 2.2).

Another measure of the influence of Wu's results is revealed by a search for the term 'Wu class' or 'Wu classes'. It returned 64 results on MathSciNet (in title or full text of the review), 679 results on Google Scholar and 25 valid results on Web of Science (WoS, in title, keywords or abstract). Comparable results for the 'Chern class', for example, are 3,189 on MathSciNet, 18,900 on

Table 2.2 Citations of Wu Wen-Tsun's, H. Cartan's and R. Thom's papers on Steenrod squares

Article	MathSciNet	Google Scholar	Web of Science
Cartan (1950)	7	27	10
Thom (1950)	5	11	6
Wu Wen-Tsun (1950a)	20	108	1
Wu Wen-Tsun (1950b)	21	72	37
Thom (1952b)	6	19	N/A
Wu Wen-Tsun (1952)	11	27	N/A
Wu Wen-Tsun (1953a)	1	3	N/A
Wu Wen-Tsun (1957c)	1	3	N/A

GS and 311 on WoS. This is because Wu classes are mainly a technical tool for calculation of other types of classes, whereas Chern classes have important applications in theoretical physics.

Topological career in China (1952–58)

This section will focus on Wu Wen-Tsun's personal experience after he arrived in Beijing in August 1951. The social and political context of Wu Wen-Tsun's work in this period are treated in more detail in the next chapter.

Wu was initially appointed full professor at Peking University. He had also been invited to work at the newly established Chinese Academy of Sciences (CAS, *Zhongguo kexueyuan* 中国科学院) by Tian Fangzeng 田方增 (born 1915), who had returned from France a year earlier. But Wu followed the advice of his French colleague Thom, who thought a university might be less stressful than a research institute.

Thom's advice turned out to be unsuitable for the Chinese situation. Wu Wen-Tsun had to teach differential geometry, a subject somewhat removed from his major interest, and felt very bad about the quality of his teaching (Zhang Zhihui, *et al.* 2008: 99). Wu was even more frustrated by the thought reform campaign aimed at 'old intellectuals' raging at the university, which involved frequent meetings with mutual criticism, including criticism of Wu's benefactor Jiang Zehan. Wu decided to accept the standing invitation to CAS, which had just formally established its Institute of Mathematics (IMCAS), and arrived there after spending a few months at a cadre training camp.[8]

At IMCAS, Wu was reunited with several familiar figures: he knew Zhang Sucheng and Zhou Shulin from the Shanghai Preparatory Office, and Tian Fangzeng and Guan Zhaozhi from Paris. Zhang Sucheng, Wu Wen-Tsun and Sun Yifeng (孙以丰, born *c*.1922, another former student of S.S. Chern) established a weekly topological seminar, which continued until 1958, when it had to be dissolved because of the Great Leap Forward. It was attended by other mathematicians as well, for example a young student of Hua Loo-Keng's theory of complex functions Lu Qikeng (陆启铿, born 1929), and sometimes even Jiang Zehan, who commuted from the relatively remote Peking University (Lu Qikeng 2010; Gan Danyan 2010).

The first six and a half years of Wu's career in IMCAS were a busy and productive period. He published at a steady pace – almost every three months. He continued to capitalize on his familiarity with characteristic classes, trying to prove for the Pontrjagin classes what he had for the Stiefel–Whitney classes, but he also developed a new interest in the study of immersion of manifolds into Euclidean space. In the mid-1950s, he briefly resumed international contacts, attending mathematical congresses in the Soviet Union, Romania, Bulgaria and East Germany, and visiting France in May 1958 (Hu Zuoxuan and Shi He 2002: 57–62). Wu was at the time one of a handful of Chinese mathematicians noted for their originality outside China. He was even invited to give a thirty-minute plenary talk at the International Congress of Mathematicians in Edinburgh in 1958, but did not attend (Stone 1961). The reasons have not been made clear, but were most probably due to the failure of China's negotiations with the International Mathematical Union about Taiwanese membership in the organization (Lehto 1998: 126–9), as well as the unfolding of the Great Leap Forward (1958–60).

In the mid-1950s, China took a series of measures to boost scientific research and raise the status of successful scientists. One of the steps in this strategy was the creation of prizes for outstanding research results. The first such prize, Science Prize of the Chinese Academy of Sciences – Natural Sciences (*Zhongguo kexueyuan kexue jiangjin* (*Ziran kexue bufen*) 中国科学院自然科学奖金 自然科学部分) was awarded on 24 January 1957.[9] Three scientists received the highest award: Hua Loo-Keng, Tsien Hsue-shen (Qian Xuesen 钱学森, 1911–2009) and Wu Wen-Tsun. Wu was the youngest and least politically engaged of the three. Wu Wen-Tsun's status as one of China's leading mathematicians was confirmed again by his election to the Academic Department of CAS in 1957 (more in Chapter 3).

In May 1953, Wu was introduced to Chen Pihe (陈丕和, four years his junior) and they married two weeks later. Chen had excellent English and typewriting skills, and by the end of 1954 was assigned a job in the library of IMCAS, where she stayed until retirement. They had three daughters and a son. Chen Pihe assumed all household responsibilities, enabling Wu Wen-Tsun to concentrate fully on his research work. She even typed many of his manuscripts for him (Hu Zuoxuan and Shi He 2002: 55).

Politically, the years 1953–8 were relatively quiet for Wu. After a major purge of a group of writers and intellectuals in 1955 (the Hu Feng incident), Wu and other mathematicians (such as Zhang Sucheng) wrote articles denouncing Hu Feng (Wu Wen-Tsun 1955a), but there are no documents showing Wu's involvement in the Anti-Rightist Campaign in the second half of 1957, which targeted outspoken, mostly young, activists. However, in 1958 he undermined his political credit during his trip to France. According to a later report, Wu Wen-Tsun 'wavered under the influence of certain French mathematicians, lost contact with China for a period, and only returned after [the CAS President] Guo Moruo urged him telegraphically', which resulted in a series of 'criticism and help meetings' he had to go through (IMCAS 1975). This murky episode probably did not cause Wu Wen-Tsun any persistent problems.[10] However, it symbolically ended the golden years of his career in close contact with frontline world research in algebraic topology.

Wu's search for a research direction

After Wu settled down in the Chinese Academy of Sciences, he devoted two years to the study of Pontrjagin classes, which had interested him since 1947 for the tough problems they presented. Several algebraic topologists (Chern, Thom, Serre, Adem, Hirsch) were trying to prove theorems about their behaviour and provide more useful definitions for them. Wu proved that certain cup-products of Pontrjagin classes are independent of the differential structure of the underlying manifold, and that these classes are topologically invariant when reduced mod odd primes (Wu Wen-Tsun 1954b) and mod 4 or 12 (Wu Wen-Tsun 1954c).[11] The latter result, together with the conjecture that the Pontrjagin class of a four-dimensional orientable manifold is three times its signature, was especially important for the proof of the generalized Riemann–Roch theorem, introduced by Hirzebruch (1956) and proved by Atiyah and Hirzebruch (1959). This later led to the proof of the Atiyah–Singer index theorem (Atiyah and Singer 1963), which is a powerful generalization of topological and analytical results with many applications in theoretical physics. Atiyah and Singer received the Abel prize for it in 2004 (The Abel Committee 2004).

Wu's five articles on Pontrjagin classes were all published in Chinese and translated into English after a shorter or longer delay. The first two (Wu Wen-Tsun 1954a, 1955b) were published in the official Chinese English-language journal *Scientia Sinica*, but the most influential (Wu Wen-Tsun 1954c) had to wait until 1959, when it appeared in the American Mathematical Society Translation Series. The last two were translated in later instalments of the same series in 1964 and 1970. In the meantime, however, Western mathematicians were already aware of the results and referred to them, e.g. John Milnor, who wrote: 'Since Wu's paper is in Chinese a proof is included in the appendix' (Milnor 1958: 444).

Ironically, the Chinese original versions of Wu's papers did not generate any citations in China at all, according to CNKI. Wu had not yet created his research school, and other Chinese mathematicians were similarly pursuing lines of research they had started earlier. Internationally, however, the first three papers (especially the last in the series) accumulated decent citation counts over the years. No citations have been found for 'Pontrjagin classes V' (Table 2.3).

Simultaneously with his work on the Pontrjagin classes, Wu Wen-Tsun probed a new research direction. Recalling this change of topic in 2004, he claimed that it was motivated by his feeling of isolation from the busy

Table 2.3 Citations of Wu Wen-Tsun's papers on the Pontrjagin classes

Article	MathSciNet	Google Scholar
Wu Wen-Tsun (1953c)	2	9
Wu Wen-Tsun (1954b)	3	9
Wu Wen-Tsun (1954c)	12	34
Wu Wen-Tsun (1955c)		4

development of algebraic topology, and search for a branch that was currently neglected in the West:

> After I returned to China in the summer of 1951, I was in a new situation, basically isolated from the outside world, or from foreign countries. I was like a solitary army preparing for war – how to do research in such a situation? I also realized that my research had been for many years limited to the cutting-edge topology of characteristic classes and fibre bundles. I thought I could perhaps expand my research range. (…) In order to solve these problems, I made an analysis and survey of trends in the history of topology.
>
> (Wu Wen-Tsun 2004: 15)

In his analysis, Wu concluded that there was a major change in topology after 1930 with the introduction of homotopy theory. Algebraic-topological invariants, introduced subsequently, were all also invariants of the homotopy type, in other words did not distinguish between spaces of the same homotopy class. Wu therefore looked for a method to construct topological invariants that preserved the finer distinctions disguised by homotopy equivalences:

> Such topological but not homotopy-type invariants are necessary, because their discovery can give us the hope of discussing topological classification of spaces, the main problem of classical topology, instead of being limited to the homotopy classification of spaces.
>
> (Wu Wen-Tsun 1953b: 261)

The method he found was based on assigning homotopy classes to collections of spaces derived from a given polyhedron K by certain constructions, featuring the p-fold topological product $K \times K \times \ldots \times K$. Although a lot was unclear about the new invariants, their variation within one homotopy type was evident. Wu's article was positively reviewed by an erstwhile colleague from Shanghai, Hu Sze-Tsen (S.T. Hu, Hu Shizhen 胡世桢, 1956), but otherwise did not generate immediate interest, and has never been translated into English.

At this time, Wu came across a paper by Egbert van Kampen (1932), as he later recalled (Wu Wen-Tsun 1965a: 284). Van Kampen proved an interesting result about the possibility of realization of abstract manifolds in Euclidean space, but the result was not connected into a single theory, and expressed in the outdated language of vector spaces. Wu wrote a bulky paper (Wu Wen-Tsun 1955c) in which he translated van Kampen's constructions into cohomology theory. This enabled him to extend the method and describe the possibility of imbedding by characteristic classes Φ^m of the non-homotopy type introduced in Wu Wen-Tsun (1953b), which he called *imbedding classes*.

One year later Wu submitted another paper (Wu Wen-Tsun 1957c), in which he noted the close relation of imbedding classes to the periodic transformations constructed by P.A. Smith (1941), and further expanded his method to the study of immersion (realization that is only locally a true imbedding) and isotopy (homotopy of imbeddings). These became objects of further developments to Wu's theory after 1959.

Some of Wu's results were independently reached by the American topologist Arnold Shapiro, who published a very similar reformulation of van Kampen's theorem (Shapiro 1957). But Wu's discovery was recognized as original in China, and he was awarded the Science Prize in January 1957 for his results:

> The work of the mathematician Wu Wen-Tsun promoted the development of one branch of geometry – topology, and has already attracted the attention of international topologists. His results concerning a basic problem of topology – the study of characteristic classes – have attained a very accomplished level. An important problem in his 'On the realization of complexes in Euclidean spaces' has not seen any outstanding results in the last twenty years. Wu Wen-Tsun used a topological property – characteristic class, integrated it with the recent development of topology, and provided a way to solve these types of problems.
>
> (Guo Moruo 1957)

By this time, Wu was finishing the third part of the series (Wu Wen-Tsun 1958b), in which he proved that the nullity of the imbedding class Φ^{2n} is both necessary and sufficient for a possible realization of the complex in Euclidean space $\mathbf{R}^{2n}$ ($n > 2$). The three papers were then collected as mimeographed notes (in English) in 1957.

Wu Wen-Tsun further intended to study imbedding of differential manifolds with the help of Whitney's theory of singularities (Whitney 1944a, 1944b). In 1958, he visited France and presented his ideas on the topic at a colloquium in Paris. André Haefliger, a Swiss mathematician attending the meeting, developed these suggestions into a series of influential articles starting with Haefliger (1961). Wu later reflected on his frustration:

> After 1960, when I could work again, I was led to these thoughts: it was me who established the theory of imbedding classes, I found a concrete method, but in the 1960s I had already fallen behind, because Haefliger

Table 2.4 Citations of selected Wu Wen-Tsun works on imbedding and immersion

Article	MathSciNet	CNKI	Google Scholar
Wu Wen-Tsun (1953b)	3	6	4
Wu Wen-Tsun (1955d)	17	7	46
Wu Wen-Tsun (1957b)	11	6	0
Wu Wen-Tsun (1957a)	3	0	6
Wu Wen-Tsun (1958b)	6	4	2
Wu Wen-Tsun (1958a)	6	0	24
Wu Wen-Tsun (1959b)	3	2	4
Wu Wen-Tsun (1960a)	9	0	15
Wu Wen-Tsun (1964b)	4	0	12
Wu Wen-Tsun (1965a)	29	0	76

had done his excellent work: if I continued to work in this direction, I would be passive; should I go on passively, or should I liberate myself from this passive situation and search for a new direction?

(Wu Wen-Tsun 2004: 17)

Nevertheless, Wu continued to publish short notes on the theory of immersion and imbedding until 1965. He also revisited his mimeographed notes, which were published in English as the book *A Theory of Imbedding, Immersion, and Isotopy of Polytopes in a Euclidean Space* by the Science Press in Beijing (Wu Wen-Tsun 1965a). The book made wide-ranging connections between Wu's theory and other aspects of algebraic topology and the history of topology; it also included Wu's first experiments with the tools of category theory and commutative diagrams, popularized in modern mathematics after the work of Eilenberg and Steenrod. Wu also adopted into his general method developments to Smith's theory of periodic transformations due to Liao Shan-Dao (S.D. Liao, Liao Shantao 廖山涛, born 1920), a student of S.S. Chern at the University of Chicago. Liao had only returned to China in 1956, having declined an offer to work at Princeton with Norman Steenrod. His papers, such as Liao Shan-Dao (1957), were rare exceptions to Wu's overwhelmingly Western citations. Liao's ideas received special praise in the introduction to Wu Wen-Tsun (1965a), as well as in the 'Historical Notes' Wu appended to the end of his book.

Wu Wen-Tsun (1955c, 1957a, 1958b) generated robust citation indexes (Table 2.4). They have also been cited by other Chinese articles, which was rare for Wu's pre-1966 works. Wu's student Yue Jingzhong (Yo Ging-Tzung) continued to work on Wu's theory (Yo Ging-Tzung 1963). All Wu's Chinese papers appeared with long English summaries and often in full translation, and his later notes were published directly in English without a Chinese version. The book that Wu eventually published in 1965 on his theory of immersion and isotopy was very successful, and remains his most-cited item in the database MathSciNet.

By 1958, China had entered a series of abrupt policy changes. Free choice of research topics, which enabled Wu to study Pontrjagin classes and the theory of immersion for several years without major interruptions, was a thing of the past. The Communist Party of China began to assume its leadership over mathematics more forcefully, and Wu's career and ideas were profoundly affected. Let us now shift our attention to the environment in which Wu was working and analyse the impact of its changes during the first 17 years of the People's Republic on Wu and his colleagues.

Notes

1 Although statistics was one of the most well-established branches of modern knowledge in China, with a national association founded in 1928 – eight years before the Chinese Mathematical Society – it was taught largely at colleges of political science and administration and not seen as essential training for natural scientists. See Mo Yueda (1993).

2 For example Soo Pu-Ching 苏步青 (Su Buqing, 1902–2003), differential geometer with a PhD from Tohoku Imperial University in Sendai, who held important academic posts both before and after 1949 (Cheng Minde 1994: 97–112); Chu Kun Ching 朱公谨 (Zhu Gongjin, 1902–61), the first Chinese PhD from Göttingen; and Chow Wei-Liang 周炜良 (W.L. Chow or Zhou Weilang, 1911–95), a student of B.L. van der Waerden in Göttingen and Leipzig, who became professor at Johns Hopkins University in 1949 and worked on differential equations, algebraic topology and geometry (Wilson, *et al.* 1996).

3 Chiang Li-fu, also known as Chiang Tsoo (Jiang Zuo 蒋佐) was one of the founders of modern mathematics in China, a PhD from Harvard (1919) and first head of department of mathematics at Nankai University in Tianjin, thus also a teacher of S.S. Chern (Cheng Minde 1994: 17–34).

4 Characteristic classes were originally a way to describe singularities of vector fields, a problem that goes back to the founder of topology, Henri Poincaré (1854–1912), and his attempts to classify ordinary differential equations (Gan Danyan 2005: 137–8). Eduard Stiefel in Zurich (1935) and Hassler Whitney at Harvard (1935) independently observed that singularities of vector fields on manifolds belong to particular topologically invariant homology classes, which they called characteristic classes. Wu Wen-Tsun became aware of characteristic classes through the work of Stiefel's supervisor Heinz Hopf.

5 Jiang Zehan was the founder of the topology school at Peking University; PhD (1930) from Harvard (Cheng Minde 1994: 113–125).

6 Norman Steenrod (1947) created a special product for a study of homotopy classes of continuous maps to arbitrary spheres. Thom and Wu brought this topic to the notice of Cartan, who gave a set of lectures on it during his 1949–50 *Topology Seminar* in Paris (Dieudonné 1989: 511). Steenrod products induce homomorphism of cohomology groups (Steenrod squares), for which Cartan formulated a set of axioms (Cartan 1950). Wu and Thom conjectured a formula which was in essence a generalization of Whitney's product theorem to Steenrod squares, and is now called the 'Cartan formula' (although Cartan acknowledged their contribution in his article).

7 It is not clear who first used this term, but it was coined sometime between Adams (1961), who preferred 'generalized Stiefel–Whitney classes', and Milnor (1965). The names 'Wu classes' and 'Wu formula' became standard through the Milnor and Stasheff (1974) textbook, where the 'Wu formula' is the subject of several exercises.

8 Wu is listed as officially entering IMCAS in October (IMCAS 1953), but Hu Zuoxuan and Shi He (2002: 52) mention December, which seems more likely.
9 The prize was worth 10,000 yuan, more than three times Wu's annual salary at the time (*Renmin ribao* 1957b). The selection process was conducted by scientists themselves, with final decisions through anonymous ballots (*Renmin ribao* 1957a).
10 Wu's reluctance to return from France in 1958 was brought up during the Cultural Revolution, in an attempt to undermine Hua Loo-Keng's position (Wang Yuan 1999: 300).
11 Pontrjagin classes are not generally topologically invariant, as proved by Milnor (1963).

3   Mathematics and Chinese socialist construction

In 1959, the People's Republic of China (PRC) celebrated its tenth anniversary. For this occasion, the Chinese Academy of Sciences (CAS) compiled a series of documents called *Chinese Science in the Past Decade*. Apart from listing new results and highlighting the quantitative growth of Chinese research infrastructure since 1949, these books also triumphantly reviewed the political and ideological transformation of each individual discipline. The volume on mathematics reflected the difficulty of this process in such an abstract field, traditionally focused on finding and proving true theorems in order to gain academic recognition. These difficulties were, however, presented as a thing of the past. Transformation of mathematics had by this time been achieved; it would now reliably fulfil its social function and assist in socialist construction, promised the preface to the volume, thanks to the enthusiastic identification with the goals of socialism by every individual mathematician:

> What shall I do in this period of astonishingly great transformation, facing the present infinitely exalting state of our country? This is a question every mathematical worker unwilling to fall behind inevitably has to consider. Searching for a way to make mathematics integrated with the production practice of socialist construction has thus become everyone's urgent requirement.
>
> (Hua Loo-Keng, *et al.* 1960: 16)

That was at the height of the Great Leap Forward (GLF), a campaign to radically increase economic output by political mobilization, which also affected science and education.[1] The disastrous results of this campaign prompted a period of Readjustment (*tiaozheng* 调整) after 1960, which was also reflected in a new, more balanced approach to theoretical science. The requirement to 'link theory with practice' (*lilun lianxi shiji* 理论联系实际), however, retained political force throughout the 1960s.

Wu Wen-Tsun's single-minded pursuit of his research interests was broken by the GLF, but his response was not stubbornly negative. On the contrary, he eventually became a vocal advocate of research that solves practical problems, and in later reminiscences traced his attention to the practical implications of mathematical research back to the experience of the GLF period. Although available documents do not explicitly reveal his contemporary views, we can attempt to reconstruct his interactions with his institutional environment and with other members of the Chinese mathematical elite. The picture that emerges from this reconstruction confirms a growing emphasis placed on practically relevant research by mathematics administrators as well as by Chinese mathematicians themselves.

This chapter focuses on the composition and development of the Institute of Mathematics of the Chinese Academy of Sciences (IMCAS) between its foundation in 1952 and the start of the Cultural Revolution in 1966. It follows two main developments: the integration of mathematical research at IMCAS with the requirements of external 'production practice', and the intrusion of politics into the daily operation of the Institute, through government planning, increased Party control and centrally orchestrated political campaigns. Wu Wen-Tsun's reaction to these developments is shown primarily through an analysis of his mathematical work in this period.

In the first section, I review existing literature on science and politics in post-1949 China. Many authors, both Chinese and Western, use metaphors such as 'fluctuations', 'oscillations' and 'twists and turns' to describe the policy environment in which Chinese scientists were working before 1978. I introduce the theoretical framework set up by Suttmeier (1974) to use it critically throughout the chapter for discussing long-term trends and goals, and short-term policy instability in Wu Wen-Tsun's research environment. Suttmeier pointed out that behind the superficial oscillations, there were stable long-term goals shared by various actors in the Chinese research sector. He explained policy instability by conscious experimentation with the means to achieve these goals. I argue that this experimentation in IMCAS reflected external political realignments and pressures, rather than urgent problems within the institute, but that these external pressures were highly internalized. Rather than portraying individual scientists as victims of thoughtless interference, I highlight the tragic element of tension between goals and results.

The second section introduces IMCAS, its organization, leading members, and its evolution in the 1950s under a relatively liberal pro-science policy. The third section discusses the impact on IMCAS of political campaigns. The fourth section analyses the Maoist requirement to 'link theory with practice', its reflection during the Great Leap Forward in Wu Wen-Tsun's and others' work, and its redefinition in debates in the 1960s. The last section introduces Wu's return to pure mathematics and his teaching at the University of Science and Technology of China (USTC). His study of algebraic geometry in the 1960s brought him into contact with concepts and techniques he later used to construct his method of mechanization.

The 'politics of mathematics' in China was not simply a struggle between the insensitive Communist Party and mathematicians trying to protect their autonomy. It was a collective enterprise with major involvement from mathematicians themselves, albeit with the Party cadres and centrally orchestrated campaigns playing decisive roles. Vague slogans were transformed into specifically mathematical requirements and imposed upon the community by its own members. Wu Wen-Tsun and other mathematicians eventually responded to these calls positively and look back on the period with a certain nostalgia.

Science and politics in China, 1949–66

The Communist Party of China (CPC) was not the first to insist on party leadership in science and technology; the Soviet Bolsheviks as well as the Chinese Guomindang had both earlier come up with similar demands. What was striking about Communist China until the late 1970s, however, was the gap separating the Party and experts. Unlike other former revolutionary parties which quickly created their own intellectuals and turned them into powerful cadres, the Chinese Communists did not trust intellectuals, including scientists.

The divide between the Party and the scientists was initially quite natural: virtually all scientists in the first years of the PRC came from relatively affluent urban middle-class backgrounds. Due to the absence of graduate education in China, almost all senior researchers had studied at foreign universities, mostly American or British; in fact the new regime actively encouraged the return of those who still remained abroad. Moreover, several scientists had some previous involvement with the old Nationalist government or its affiliated organizations. In short, their degree of identification with the new regime was considered low, and their understanding of its goals insufficient (Yao 1989).

But unlike in other totalitarian regimes, even CPC's reliable supporters among the scientists did not quickly rise to positions of power. The Chinese Communists harboured a profound suspicion of all experts, grouped together with scholars in the humanities, artists and writers into the category of intellectuals (*zhishifenzi* 知识分子, literally 'knowledge [possessing] elements'). Intellectuals stood low in the class hierarchy and were from the outset subject to repeated thought reform campaigns, being considered permanently susceptible to dangerous individualism and liberalism (Goldman 1987a: 236).

On the other hand, science was crucial for the CPC in two equally important respects. Its development was necessary for the modernization project, to which the CPC unreservedly subscribed, and for the construction of a strong and independent country (Teiwes 1987). The CPC's claim to broad legitimacy was staked (among other things) on making the new regime better for the advancement of science, and especially for its application to national economy and defense, than the Kuomintang (KMT)-run Republic. Science was also the basis of Communist ideology, and science dissemination was thus part of ideological indoctrination, especially for the suppression of traditionalist rival ideologies among the masses (Suttmeier 1970).

Scientists thus had to be kept active if these hopes were to be realized. And their expert advice was needed for crucial decisions on strategic technologies and allocation of research funding. The central theme of the 1949–66 period is therefore a tension between science and politics (exacerbated during the Cultural Revolution). Because scientists combined the identity of technical experts and of liberal intellectuals, changing policies towards intellectuals were reflected, albeit in a milder manner, in the conditions under which they worked.

Existing literature on China's post-1949 science treats these themes in various ways. Institutional histories compiled in the PRC usually downplay political instability and portray the development of science policy as driven by science-internal rational considerations (Chen Jianxin, *et al.* 1994). On the other hand, accounts focused on individuals or disciplines often treat politics as a purely external influence, the 'bigger climate' (*da qihou* 大气候), which cannot be properly explained. Virtually all literature on the history of modern mathematics is written in this way, including the biography of Hua Loo-Keng, which is at the same time the most extensive history of IMCAS so far (Wang Yuan 1999). Wu Wen-Tsun's biographies (Hu Zuoxuan and Shi He 2002; Ke Linjuan 2009), as well as the historical survey of modern Chinese mathematics (Zhang Dianzhou 1999), also treat politics as an external, non-negotiable factor.

Accounts written by Western scholars mostly emphasize the antagonistic division between the Party and the intellectuals, focusing especially on the fate of the older Western-trained scientists (Goldman 1987a, 1987b; Yao 1989; Cao 1999). This is a well-established perspective on PRC history, embodied especially in the *Cambridge History of China* (MacFarquhar and Fairbank 1987) and in the other writings of Roderick MacFarquhar (MacFarquhar 1983, 1997; MacFarquhar and Schoenhals 2006). A recent study of engineers translates this division into the competition between political and cultural capital (Andreas 2009). Experts accumulated cultural capital, which was needed for the modernization and socialist transformation of Chinese economy. They expected a corresponding share of political power, but in China, unlike in Soviet-type Communist societies, such a technocratic transformation did not happen before 1976, because it ran against the egalitarian, anti-elitist core of Mao Zedong's beliefs. Nevertheless, the Party remained ambiguous about the role of experts and kept nurturing their technocratic ambitions.

However, this focus on the control over intellectuals suppresses the realistic, pragmatic motivations of the CPC's steering of the Chinese science and research sectors. These were elaborated (for the period 1949–70) in Richard Suttmeier's book *Research and Revolution*. Suttmeier speculates that the often abrupt policy changes were intended to overcome imbalances in the science and technology sector and unblock its further development (Suttmeier 1974: 29–33). In Suttmeier's view, the Party had a set of long-term goals for the science and policy sector, which could not be pursued all at once. They were therefore approached separately in consecutive 'states of regularization' and 'states of mobilization'. In the states of regularization, research and education were run in an orderly but also increasingly bureaucratic manner, which caused loss of enthusiasm and waste of resources. When these problems

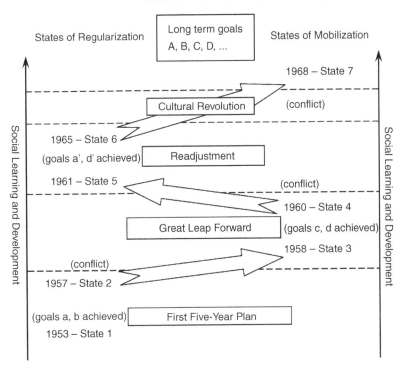

Figure 3.1 'Social experimentation' and oscillations in the Chinese research sector, 1953–68

Source: Based on Suttmeier (1974: 31, fig. 2-2).

became obvious, the Party initiated a political campaign which ushered in a 'state of mobilization'. Power shifted from senior experts to revolutionary cadres and activists. This allowed a rapid change of direction and the application of the resources gathered during the preceding stage to new goals. When this eventually exhausted the system, experts were called back to start a new state of regularization (see Figure 3.1).

Suttmeier's model gives meaning to the frequently observed pattern of policy 'oscillations' or 'fluctuations' since 1949.[2] Figure 3.1 suggests that once development within a 'state of regularization' reached a critical point (e.g. State 2 or State 6), a period of conflict ensued (indicated by dashed lines), in which decision-making and overall orientation was changed to a 'state of mobilization'. Similarly, when mobilized development reached a critical stage at the end of the Great Leap Forward, a period of conflict brought the system back into a more regular state with firmer bureaucratic control. But these oscillations were not cyclical, they drove the system forward towards long-term goals along a path of 'social learning and development'.

Suttmeier's theory is obviously very schematic and also rather naïve in its explanation of all political campaigns by conscious goal-oriented experimentation.[3] But its two most original elements can usefully frame my description of IMCAS and Wu Wen-Tsun in this chapter.[4] Suttmeier's model emphasizes long-term continuities and shared visions behind the apparent fluctuations of actual policy, and he attempted to explain these changes by factors within the science and research sector itself, rather than as mere reflections of overall political shifts in China to the left or right.

As I will show, these points are to some extent valid for the history of Chinese mathematics between 1949 and 1966, more specifically for the history of IMCAS. During the entire 17-year period between the establishment of the PRC and the Cultural Revolution, including the Great Leap Forward and the post-GLF Readjustment, a planned extensive and intensive growth of Chinese mathematics was indeed a shared long-term goal, pursued by all leaders of IMCAS. Moreover, although the timing of policy shifts was certainly determined externally by centrally orchestrated campaigns, the actual form of these policy shifts within Chinese mathematics and IMCAS was local, and thus reflected genuine problems as perceived and articulated by mathematicians. Even on the individual level – illustrated here by Wu Wen-Tsun and his colleagues – the policy shifts were met with adjustments to one's research plans in line with certain long-term goals, and were seen as an opportunity to escape problems created by previous concentration on one research area. Without denying the disruptive effects of this instability, we can, using Suttmeier's framework, better understand how it gradually transformed Wu Wen-Tsun and his coworkers.

This approach to the history of PRC science is in line with recent trends to overcome the discourse of disruption and damage, and view the actual course it has taken as an organic result of the interactions of Chinese scientists with their social environment.[5] This chapter attempts to show the ways in which the form and content of mathematics produced in the PRC was shaped by the social environment, by popular and political demand, and to pinpoint where these transformations left Wu Wen-Tsun at the start of the Cultural Revolution.

The Institute of Mathematics, Chinese Academy of Sciences

In 1949, the new communist state inherited a modest number of scientists and research institutions from its predecessor, the Republic of China (Zhang Dianzhou 1999: 125–75). As already mentioned, most of them had studied at foreign universities, and often continued to pursue questions investigated at these scholarly centres.

During the first three years of the PRC, the most prominent mathematicians resident in China set up a Preparatory Office for the Institute of Mathematics (*Shuxue yanjiusuo choubeichu* 数学研究所筹备处) in the Chinese Academy of Sciences in July 1950, and actively encouraged mathematicians studying or working overseas to return home. The Preparatory Institute was first headed by the Shanghai-based geometer Su Buqing (苏步青, 1902–2003), but he relinquished the post in favour of Hua Loo-Keng, an ambitious mathematician who had just returned from the USA in March 1950. The Institute of Mathematics was formally established, with Hua as director, on 1 July 1952.[6]

The Chinese Academy of Sciences (CAS, *Zhongguo Kexueyuan* 中国科学院) was initially a government department founded on 1 November 1949, exactly one month after the proclamation of the People's Republic. It incorporated all research institutions that remained on the Chinese mainland at that time, including major parts of the two earlier government-sponsored academies, Academia Sinica (*Guoli Zhongyang Yanjiuyuan* 国立中央研究院) and Peiping Academy (*Beiping Yanjiuyuan* 北平研究院).

CAS exercised administrative control over its research institutes (*yanjiusuo* 研究所) as well as 'academic leadership' over the entire Chinese science sector. Administrative decisions were taken by the CAS Secretariat, staffed by CPC cadres. To boost its academic leadership, CAS established Academic Departments (*xuebu* 学部, AD) in June 1955 on the model of the Soviet Academy of Sciences, following a long process of selection of the most prominent scholars in each discipline. Membership in an Academic Department was a position similar to academicians in pre-1949 China and in the Soviet Union, but the term 'academician', and the high social status that accompanied it, were not officially awarded to the members, as the Party feared it would elevate non-party scientists too much (Cao Cong 1999).

IMCAS director Hua was one of these influential non-party scientists, although in other ways he was not a typical member of the Chinese scientific elite. He came from a rather poor merchant family and had no formal education beyond junior high school. He gained his mathematical training as a non-qualified assistant at the Department of Mathematics at Tsinghua University in the 1930s, and thanks to his talent and hard work gradually proceeded through a two-year research stay at Cambridge, professorship at Tsinghua University (1937–43), fellowship at Princeton and finally a professorship at the University of Illinois. He also visited the USSR in 1946 and, most importantly, decided to return to China in 1950. He vocally supported the new regime in an 'Open Letter to Fellow Chinese Still Abroad', which was widely publicized by the *New China* press agency. Hua urged his colleagues abroad to return to China and help in its development. The letter rejected Western freedom and democracy as 'mere words', and emphasized:

> There are two camps in the current world. In one camp the common good is sought for the masses, while the other camp specializes in keeping most of the benefits for the small number of people belonging to the ruling class.

> (Wang Yuan 1999: 152)

In this way, he became a public face for the new regime. He also held important political posts, including membership of the Standing Committee of the National People's Congress (a showcase parliament without effective power)

Table 3.1 Senior mathematicians working at IMCAS before 1964

Research Professors (RP), Members of the Academic Department (AD)	
Hua Loo-Keng [Hua Luogeng] 华罗庚 1910–85	RP & Director 1950. AD 1955. No formal degree. Number theory, abstract algebra, complex functions. Cambridge 1936–7 (G.A. Hardy). Professor at Tsinghua and Southwest Associated University (Kunming) 1937–46, University of Illinois 1947–50. China Democratic League (Vice-Chairman), Standing Committee of the National People's Congress (1956). (Wang Yuan 1999)
Wu Wen-Tsun [Wu Wenjun] 吴文俊 Born 1919	RP 1952. AD 1957. Student of S.S. Chern. PhD Strasbourg 1947–9 (C. Ehresmann). Algebraic topology. Research fellow CNRS Paris 1949–51. Professor at Peking University 1951–2. (Hu Zuoxuan and Shi He 2002: 60)
Chang Tsung-Sui [**Zhang Zongsui**] 张宗燧 1915–69	RP 1956. AD 1957. PhD Cambridge 1936–8 (R.H.Fowler). Theoretical physics. Research in Copenhagen (N. Bohr), Zurich (W. Pauli), Cambridge (M. Dirac), Princeton. Professor at National Central University (Chongqing) 1939–45, Peking University, Beijing Normal University 1951–6. (Cheng Minde 2000: 206–18)

Research Professors (RP)	
Ou Sing-Mo [**Wu Xinmou**] 吴新谋 1910–89	RP 1951. Paris (J. Hadamard, PhD?). Partial differential equations. Research in France 1937–51. CPC 1945. French wife, both members of French Communist Party. (Cheng Minde 1994: 189–98)
Ming Nai-Ta [**Min Naida**] 闵乃大 Dates N/A.	RP 1952. PhD Leipzig 1939, TU Berlin 1944 (W. Cauer). Theoretical electronics. Tsinghua University 1948–52. German wife. Emigrated to East Germany 1958, then West Germany. (Xu Zuzhe 2010)
Hu Shih-Hua [**Hu Shihua**] 胡世华 1912–98	RP 1953–61 (to Institute of Computational Science). PhD Münster (Heinrich Scholz, G. Köthe). Mathematical logic. Professor at Zhongshan University (Guangzhou) 1941–3, National Central (Chongqing) 1943–5, Peking University 1946–53. CPC 1954. (Cheng Minde 2000: 164–84; Lu Jiaxi 1992: 33–7)
Chang Su-Cheng [**Zhang Sucheng**] 张素诚 1916–2006	ARP 1950, RP 1953. Student of S.S. Chern. DPhil Oxford (J.H.C. Whitehead). Algebraic topology. Professor at Nanchang University 1950, Zhejiang University 1950–3. CPC 1960. (Cheng Minde 1998: 225–42)
Hiong King-Lai [Xiong Qinglai] 熊庆来 1893–69	RP 1957. Studied in Paris, Montpellier and Grenoble 1915–20, PhD Institut H. Poincaré 1933. Function theory. Professor at Southeast University (Nanjing), Tsinghua 1928–31, 1933–7, President of Yunnan University 1937–49. Private researcher in Paris 1949–57. Close links with the Nationalist Party (KMT). (Cheng Minde 1994: 35–47)

Table 3.1 (cont.)

Research Professors (RP), Members of the Academic Department (AD)

Sun Keding 孙克定 1909–2007	RP 1958. Graduated from SNUC. Calculating devices and practical mathematics. Served in political and administrative positions in CPC's New Fourth Army, etc. Deputy Director of Zijinshan Astronomical Observatory. CPC 1930s. (Zhou Ping n.d.; Guo Lei and Wang Yuefei 2007; Zhu Guangyuan and Li Xin 2005)
Kwan Chao-chih **[Guan Zhaozhi]** 关肇直 1919–82	ARP 1952, RP 1959. Studied in Paris (M. Fréchet, no degree). Applied analysis, functional theory, control theory. Lecturer Yen-ching University 1941–6, Peking University 1946–7. CAS Translation Bureau 1949–52. CPC 1946, CPC CAS Committee. (Cheng Minde 1994: 366–76)
Ye Shuwu 叶述武 1911–96	RP 1961. Master studies in Lyon 1935–7. Theoretical physics, celestial mechanics, space programme. Zhongshan University, Zijinshan Observatory, Institute of Mechanics of CAS. CPC. (Cheng Minde 1995: 196–205; Jing Bao 2010)

Associate Research Professors

Tian Fangzeng 田方增 Born 1915	ARP 1950. PhD Paris 1947–50. Functional analysis. Lecturer at Southwest Associated University (SAU) and Tsinghua. CPC in 1938, re-entered 1956. (Cheng Minde 2000: 246–255)
Wang Shouren 王寿仁 1916–2009	ARP 1950. Studied in China (Peking University, Southwest Associated University). Probability and statistics. Secretary and Lecturer at SAU and PKU. China Democratic League. (IMCAS 1958c; Cheng Minde 2000: 265–73)
Yue Minyi 越民义 Born 1921	Research Trainee 1951, ARP 1956. Studied in China (Zhejiang University in wartime Zunyi and Meitan). Analysis, operations research. Hua Loo-Keng's early assistant. (Cheng Minde 2002: 246–53)
Qin Yuanxun 秦元勋 1923–2008	ARP 1956. MA, PhD Harvard 1944–7. Analysis, ordinary differential equations. Cadre in Yunnan 1950–3, CAS 1953–6. CPC 1959. Ordinary differential equations. Military research after 1960. (Du Songzhu and Zhang Suochun 2000; Cheng Minde 1994: 457–71)
Xu Haijin 许海津 1916–?	ARP 1958. Studied and worked in the USA 1947–58. Control theory. Persecuted in the Cultural Revolution. (IMCAS 1958c, 1972b)
Xu Guozhi 许国志 1919–2001	ARP 1960. PhD Kansas (1954, G.B. Price). Analysis, control theory. Institute of Mechanics CAS 1956–60. (Cheng Minde 2000: 294–313)

Sources: As indicated. Life dates updated according to news reports and personal communication. Bold type indicates the form of name generally used in this book.

and the Vice-chairman of the China Democratic League (*Zhongguo minzhu tongmeng* 中国民主同盟), one of the officially allowed non-Communist parties under the leadership of the CPC.

But several senior members of IMCAS (Research Professors or Associate Research Professors) had stronger links to CPC than Hua and eventually joined the Party (Table 3.1).

All senior members had studied in Western Europe or America. Those with French experience were especially numerous, and three of them – Wu Xinmou, Guan Zhaozhi and Tian Fangzeng – held important positions in the IMCAS administration.

As well as Research Professors (*yanjiuyuan* 研究员) and Associate Research Professors (*fu yanjiuyuan* 副研究员), there were also two junior ranks in CAS: Assistant Research Professors (*zhuli yanjiuyuan* 助理研究员) and Research Trainees (*yanjiu shixi yuan* 研究实习员). There was also at times a small number of graduate students (*yanjiusheng* 研究生). The number of research trainees expanded much more rapidly than that of full and associate research professors (Table 3.2).

After some promotions in the 1950s, the higher ranks became almost closed between 1956 and 1964. During the Great Leap Forward, the Institute temporarily accommodated up to 100 visitors from universities around the country. The number of permanent research trainees also exploded during this period, and there were even plans to increase the number of researchers by a hundred or more every year (IMCAS 1960b). However, many of the new trainees performed very disappointingly and had to leave the Institute in 1961–2. Later, recruitment proceeded more carefully, mostly from CAS's own source of high-quality graduates, the University of Science and Technology of China.

For the most part of the 1950s, Hua Loo-Keng was able to imprint his will on the organization of IMCAS without much political intrusion. He was by all accounts a competent and confident organizer (Wang Yuan 1999; Halberstam 1986) and tried to establish and secure IMCAS's position as a central node of the Chinese mathematical community, important for the development of Chinese science and technology in general. The institute started to organize weekly seminars in linear algebra and group theory, number theory (both headed by Hua himself), differential equations (headed by Wu Xinmou), functional analysis (headed by Tian Fangzeng), topology (with Wu Wen-Tsun, Zhang Sucheng and Sun Yifeng, all students of S.S. Chern) and Fourier integrals (later expanded to function theory, headed by Hua). These seminars were frequented by interested mathematicians from other Beijing academic institutions. In 1956 and 1957, a seminar every Friday featured IMCAS as well as non-IMCAS speakers, including five talks by an Indian mathematician D.D. Kosambi (*Shuxue jinzhan* 1957). Hua and his colleagues also particpated at nationwide mathematical conferences organized by the Chinese Mathematical Society (CMS). The first Chinese mathematical conference in September 1953 included 15 reports and 17 research papers, mostly on analysis, which was promoted because of its high level of development in the Soviet Union. In August 1956, another conference was held in Beijing at

Table 3.2 Research staff at IMCAS, 1952–66, by rank/category

	Research Professors					Total
	Full	*Assoc.*	*Assist.*	*Trainees*	*Students*	
Nov 1952	4	3	7	13		27
Dec 1954	5	5	6	13		29
Dec 1956	6	6	12	49	3	76
Dec 1958	9	6	10	50	3	78
Dec 1960	10	6	12	110	3	141
Oct 1962	9	6	23	115	6	159
Dec 1964	8	13	25	147	13	206
Jun 1966	8	13	28	167	15	231

Sources: IMCAS (1952a, 1955, 1957, 1959b, 1960b, 1962a, 1964, 1966a).

which more than 170 papers were presented, the largest such meeting before the 1980s (Ren Nanheng and Zhang Youyu 1995: 199–217).

Hua's aim was to continue to develop the parts of pure mathematics which had reached an advanced level in pre-1949 China, and at the same time to open up new branches of both pure and applied mathematics which had been ignored or neglected. He summed up his vision in his talk at the 1956 conference:

> When we talk about reaching the advanced world level, we have to arrive at a common understanding of what such a level means. We believe that if there are only individual or very few prominent mathematicians who are internationally known, such as in our current situation, it cannot be called reaching the advanced world level. Needless to say, if there is a lot of work being done, but its quality is low and its results trivial, this cannot be called reaching the advanced world level either. Even if both the quality and the quantity of work are fine, but the direction of development is too narrow, this will not suit such a large country as ours.
>
> (Ren Nanheng and Zhang Youyu 1995: 211)

Already in 1952, Hua Loo-Keng had edited a report about the future direction of IMCAS, which eloquently argued that only a broad-based development of mathematics could provide for all its necessary roles in China's socialist construction. The report criticized the excessively foreign orientation of pre-1949 Chinese research, in terms of topics, theories and publication channels, the result of which was, among other problems, an extremely narrow specialization – the worst example being the old Institute of Mathematics, which under S.S. Chern had focused almost entirely on algebraic topology. The report countered this with a compelling metaphor:

> Science is not a string of pearls – pearls to be admired by the leisurely classes; it is a machine in which one gear fits into another – a machine serving the people. Incapacity of any gear will affect the

entire machine, either decreasing its efficiency or blocking it entirely. Therefore, scientific development must be universal, so that every discipline can realize its gear-like function. Just like every gear has its neighbouring gears, every science also has its neighbouring sciences, has an aspect dealing with principles, as well as an aspect closer to applications. (...) Therefore at the present stage, our institute should pay attention both to theoretical and applied work. Each employee should extend his efforts on the topic he excels in, but also keep in mind the neighbouring branches.

(IMCAS 1952b)

The report suggested expanding theoretical subjects 'so that there are people working on algebra, geometry and analysis, and in future there is even more specialization, so that there are people working in even more narrowly classified subjects, and in the more distant future, there is a specialist on every major branch of theoretical mathematics in China.' In applied mathematics, it was necessary to fill major blanks, but also to avoid repetition of work pursued elsewhere.

Hua Loo-Keng had emphasized the importance of applications of mathematics even during the Republican period (Wang Yuan 1999: 121), and actively promoted new branches of applied mathematics in IMCAS – computational mathematics (developed by Min Naida and later Feng Kang) and mechanics in particular. Mechanics was a target of concentrated government attention because of its use for the development of ballistic missiles. The mechanics group in IMCAS included the rocket scientist Tsien Hsue-shen (Qian Xuesen), who returned from the USA in 1955 and became Director of the Institute of Mechanics, which designed China's first rockets.

Hua was also known for his insistence on practical calculation techniques. He wrote in 1944:

As far as the content of a university mathematics course is concerned, there is too much emphasis on the abstract, with concrete matters being neglected and numerical calculations often dismissed as being easy or trivial. Thus generations of students go through the system without coping with anything set in a concrete context.

(Wang Yuan 1999: 169)

Technical mastery of calculation was a dominant feature of Hua's own research in analytic number theory.[7] He attracted high-school students to this field in his promotional activities for the Chinese Mathematical Competition (Hua Loo-Keng 1956), and more advanced mathematicians at the 1956 mathematical conference by concluding his report on open problems with the sentence: 'Anyone who can make even a tiny improvement or advancement on these problems will be regarded as a world record holder' (Wang Yuan 1999: 203).

Table 3.3 IMCAS research groups and their sizes in 1953 and 1958

Group name	Members		Group name	Members	
	1953	*1958*		*1953*	*1958*
Partial differential equations	—	31	Ordinary differential equations	5	15
Functional analysis	3	14	Probability and statistics	3	22
Mathematical logic	1	16	Algebra	2	22
Number theory	6	7	Function theory	2	5
Geometry and topology	3	15	Mathematical physics	2	6
Applied mathematics	1	3	Computation science	1	Separate institute

Sources: IMCAS (1953, 1958c).

But other disciplines expanded more rapidly in the 1950s: differential equations, probability and statistics, and computation science, which eventually gained an institute of its own. Table 3.3 shows the growth of research groups established in 1953 and transformed into research sections (*yanjiushi* 研究室) in 1958. Differential equations benefitted from the most support, as a subject advanced in the Soviet Union and important in science and engineering applications.

Some theoretical subjects (algebra, geometry and topology) also did well, but number theory was not so successful. It was widely perceived as the discipline at the extreme of impracticality. Even Hua almost apologized for his research interests when he received the Science Prize in 1957[8] (for his work on function theory rather than number theory):

> I thoroughly understand that this prize is not an award for my research direction. (...) When we consider that computation science is only being created from scratch, that the number of workers in probability theory and mathematical statistics still lags far behind our country's needs, that differential equations are only very arduously rising to a higher qualitative level, we have the duty to wish that our youth studies these aspects urgently needed by our motherland. We should not think that the direction for which we received the prize is a direction the country needs the most.
>
> (Hua Loo-Keng 1957)

Available documents show that Hua's direction for Chinese mathematics was not challenged in the 1950s. His future major opponent, Guan Zhaozhi, wrote a memorandum to the head of the CPC Propaganda Department, Lu

Dingyi, in July 1956, to point out shortcomings in China's scientific organization. Guan mentioned 'old intellectuals, among whom excessive pride and haughtiness are endemic', but his criticism mostly targeted people who had entered the research sector since 1949 purely on the basis of their administrative abilities. In Guan's assessment, they often excused their academic incompetence by their administrative burden. Therefore, 'what often appears to be only a struggle between [politically] progressive and backward [elements], is in fact (or at least partly includes) struggle between people who are afraid to do research, and those who promote it' (Guan Zhaozhi 1956d).

Guan's article on the philosophy of mathematics (Guan Zhaozhi 1955) also voices a concern for developing both applications and theory, similar to Hua Loo-Keng's views. He used his considerable knowledge of mathematics and ability to read Russian and German to freely and creatively apply principles of Marxism–Leninism to mathematics. He warned against the misleading effects of idealist mathematical philosophy on mathematical practice:

> Mathematics is a natural science distinguished by its highly abstract character. This feature is often exaggerated and misused by idealists to defend their absurd teachings. From Pythagoras and Plato in ancient Greece up to the followers of Ernst Mach and logical positivists in modern time, it has always been so. In capitalist countries, many mathematicians under the influence of their environment succumb to distortion of their own scientific contributions by following the idealist philosophers.
>
> (Guan Zhaozhi 1955: 36)

According to Guan, steering clear of idealism should not mean avoiding philosophy of mathematics altogether. That would amount to irresponsibly 'leaving the battleground to the adversary'. Moreover, Chinese mathematicians' lack of confidence in philosophy made them avoid promising disciplines such as mathematical logic and finitist mathematics, which had previously been linked to 'idealist' philosophies such as formalism or intuitionism. Guan demonstrated in his article how Soviet mathematicians successfully created new mathematics by boldly putting techniques and concepts born of these idealist philosophies on a firm 'materialist' footing.

Guan used a relatively obscure, broad definition of idealism from Lenin's *Conspectus of Hegel's book 'The Science of Logic'* as 'a *one-sided*, exaggerated (…) development (inflation, distension) of one of the features, aspects, facets of knowledge, into an absolute, *divorced* from matter' (Lenin [1915] 1976, emphasis original). Such an exaggeration can occur both ways, as Guan illustrated: pure mathematicians exaggerate the role of axiomatic systems, engineers of empirical formulae; some teachers neglect computational skills, other analytical skills; some 'idealists' expect every aspect of knowledge to be quantifiable, while others deny even the legitimate uses of mathematics in natural and social sciences. Overall, the point of Guan's article was to overcome ideological narrowness of mathematical research, in contrast to his later firmly ideological stance.

The orderly growth of IMCAS culminated in the years 1955–7. Several promising developments occurred or produced their effects in this period. The Chinese government commissioned the *Twelve-Year Plan of Development of Science and Technology* (1956–67), later remembered with nostalgia as the ideal combination of science and democracy, because of the careful consultations involved in its preparation (Hu Weijia 2005; Yang Wenli and Zhang Meng 2007). The main spirit of the plan was a 'warp and weft' structuring of the Chinese research sector, primarily along projects and tasks ('the warp') to give impulses to scholarly disciplines ('the weft'). This organizing principle was codenamed 'tasks driving the disciplines' (*renwu daidong xueke* 任务带动学科). The planning committee chose the 12 most pressing tasks and four key technologies, but also added a vague category 'basic theoretical problems in natural sciences' at the suggestion of the Premier Zhou Enlai. Hua Loo-Keng represented mathematics on the planning committee (Wang Yangzong and Cao Xiaoye 2010: 81) and made sure that it was not left behind.

In 1956, the Premier also publicly declared the natural sciences a priority of the next five-year plan, promising a major expansion of funding and personnel resources committed to them in his speech 'March on Science' (*xiang kexue jin jun* 向科学进军). Salaries were raised and better housing and office space allocated to CAS institutes (Wang Yangzong and Cao Xiaoye 2010). The first award of the CAS Science Prize in 1957 was also a result of this new strategy (although it had been in preparation since 1955).

Some of the most highly regarded Chinese scientists still working or studying in the Western countries returned in 1955–7, such as the mechanics expert Tsien Hsue-shen, the function theorist Hiong King-Lai (Xiong Qinglai), and the algebraic topologist S.D. Liao (Liao Shantao, professor at Peking University), whose work was cited on several occasions by Wu Wen-Tsun. A first attempt was made at creating regular postgraduate training in China on the Soviet model of 'Candidates of Sciences', although the two years recruited (1955 and 1956) did not finish their courses and never received any titles (Gan Danyan 2010). IMCAS established an academic committee in 1955 with a majority of senior non-CAS mathematicians, but also including Hua Loo-Keng and Wu Wen-Tsun (Wang Yangzong and Cao Xiaoye 2010). In 1956, two relatively well-educated and liberal cadres Zhang Jingfu (张劲夫, born 1914) and Du Runsheng (杜润生, born 1913) were appointed to head the CAS Secretariat, also a move accommodating scientists' concerns.[9]

During the 1950s, IMCAS contributed a growing number of articles to the only dedicated mathematical research journal in China in this period, *Acta Mathematica Sinica* (*Shuxue xuebao* 数学学报). As Figure 3.2 shows, IMCAS was also more resilient to adverse conditions during the economic crisis of 1960–1, when comparatively little was produced by non-IMCAS mathematicians.

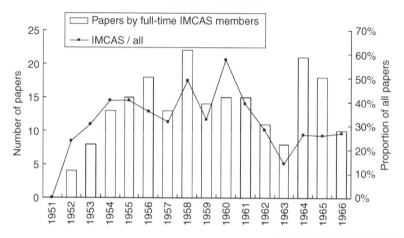

Figure 3.2 Papers from IMCAS published in *Acta Mathematica Sinica*, 1951–66

In the mid-1950s, IMCAS benefited from contacts with foreign, mainly Eastern European, mathematicians. Pál Turan from Hungary and Heinrich Grell from East Germany visited in 1954, I.G. Petrovskiy from the USSR and K. Kuratowski from Poland in 1955, and D.D. Kosambi from India and A.V. Bitsadze from the USSR came in 1957 (Wan Zhexian 1955; *Shuxue jinzhan* 1955, 1957; Yang Le and Li Zhong 1996; IMCAS PDE Group 1958). Chinese delegations attended mathematical congresses in Bulgaria in 1954 and 1956, Poland in 1955 (Wu Xinmou 1956), Romania in 1956 (Chen Jiangong, *et al.* 1956), and the Soviet Union (Guan Zhaozhi 1956b, 1956a). Contacts with Japan were re-established in 1955 by a delegation led by Su Buqing, who had studied there (*Shuxue jinzhan* 1956).

This state of 'regularization' came to an abrupt end, however, for reasons that can hardly be seen as internal to mathematics. On the other hand, as we shall see in the next section, the dynamic of this change was strictly within the mathematical community.

Political campaigns in the 1950s

[In 1957–9], the Party strengthened its leadership of mathematical activities. A thorough condemnation was made of the view expressed by many in the past, that specific tasks in mathematical research and teaching can only be led by experts. The broad masses of the mathematical community accepted Party leadership and realized that mathematical activities not only should, but also can be led by the Party, and that the Party leads concrete activities very well.

(Hua Loo-Keng, *et al.* 1960: 16)

Politics and ideology played a role in IMCAS even in the relatively quiet part of the 1950s. Sun Yifeng, Wu's colleague and also a former student of S.S. Chern, left IMCAS in 1955 to avoid Guan Zhaozhi's incessant questioning of his links to a distant uncle in Taiwan.[10] But Guan Zhaozhi's position was also far less secure than his CPC membership might suggest. According to an evaluation report compiled partly by him and partly by the CPC branch in IMCAS in the 1960s, Guan's sponsors who had introduced him to the Party were purged in the 1950s, which made him aware of his own ideological deficiencies (Guan Zhaozhi and IMCAS 1964).

In 1956, Mao Zedong initiated the 'Hundred Flowers' policy (also called 'Blooming and Contending', because of the main motto 'Let hundred flowers bloom and hundred schools of thought contend' [*bai hua qi fang, bai jia zheng ming* 百花齐放，百家争鸣]; see Goldman 1987a: 242–8), followed by a CPC rectification campaign. The masses, especially intellectuals, were asked to express their views and criticize the shortcomings of their superiors. By the spring of 1957, the criticism had become too bold and started to implicitly challenge even the principle of Party leadership. Mao was persuaded by his anxious colleagues on the Politburo to halt the rectification and instead launched a campaign to identify and isolate the 'rightists' (the Anti-Rightist Campaign). These were in the first instance the most vocal critics from the preceding period, but eventually almost anyone not in favour with the leaders of their work unit could become a victim.[11]

The campaign struck different mathematicians in different ways. It was quite ruthless at many universities, where young students led the attack on their teachers or outspoken colleagues.[12] The most serious cases were punished by reassignment to villages or factories, where they were effectively deprived of personal freedom. Many more were simply criticized at public meetings and had to go through a stressful exercise of writing self-critical confessions. The Chinese Academy of Sciences was to some extent protected. The CAS Secretary Zhang Jingfu personally requested Mao Zedong to spare senior scientists, many of whom had returned to China from abroad just a few years before the Anti-Rightist Campaign (Liu Zhenkun 1999b). But younger scientists were often affected, including five IMCAS members who were temporarily 'sent down for manual labour' (*xiafang laodong* 下放劳动, a form of forced internal exile).[13]

The director Hua Loo-Keng also only narrowly escaped a rightist label. He had participated in an initiative of fellow members of the Academic Departments of CAS to suggest improvements to Chinese science and technology policy and governance (Zeng Zhaolun, *et al.* 1957). Although it was solicited by Premier Zhou Enlai and included only mild criticism of the insufficient attention and protection given to scientists, it was published on 9 June 1957, the day after the CPC collectively decided to stop the rectification campaign and attack the rightists who had exposed their views in it (Wang Yuan 1999: 220–5).

Hua was not officially labelled a rightist, partly because he was quick to distance himself from the fateful text and partly because all natural scientists among the authors were treated leniently (Yao 1989). But his position was now less secure. This had profound implications in the subsequent campaigns of 1958.

Guan Zhaozhi also faced criticism in 1957–8 for 'the residues of old intellectuals' thought' and was reminded that because he had not participated in the revolutionary war, he had to prove himself in the current campaigns (Guan Zhaozhi and IMCAS 1964). He started writing much more aggressively in 1958.

At the same time as the Anti-Rightist Campaign was winding down, the Chinese leadership launched a new economic strategy, the Great Leap Forward (GLF, *Da yue jin* 大跃进). In February 1958, intellectuals were urged to 'alter their pace rapidly in order to meet the high tide of culture' (*Renmin ribao* 1958). The recent suppression of dissent had removed prudent voices which might speak against the inflated enthusiasm coming from all sides.

The IMCAS Party Committee (rather than the humiliated and insecure Hua Loo-Keng) submitted a GLF plan in May 1958, but over the summer changes were not very radical (Wang Yuan 1999: 233). The question of the proper direction for the development of Chinese mathematics was, however, discussed very actively. The crucial contentious issue was what counted as the correct implementation of the requirement to 'link theory with practice' (*lilun lianxi shiji* 理论联系实际). This had always been an officially promoted development goal, theoretically based on Mao Zedong's writings (Mao Zedong [1937] 1960), and clearly put forward in a speech of Mao's chief ideologue Chen Boda, Vice-President of CAS, delivered to CAS and other scientists in 1952 (Ch'en Po-ta 1952; Suttmeier 1970). The slogan meant that theory must be connected to reality in an epistemic as well as practical sense. Theory divorced from practice was bad not just because it hampered national economic construction; more importantly, it was inevitably idealist and thus ultimately false. The slogan combined ideology with science policy and a 'developmental direction' (Li Zhenzhen 2002). As long as these could be kept separate, a broad interpretation of linking with practice was possible. GLF sources revisit this common understanding of 'linking with practice' before the Great Leap Forward:

> In this period, the two directives of 'research according to plan' and 'linking theory with practice' were still quite unfamiliar to many mathematicians. There were still many who thought that 'as scientific research is an inquiry into the unknown, one cannot set a plan for it.' As for linking theory with practice, mathematicians had already become aware of this problem, but they did not grasp it profoundly and did not generally pay enough attention to it; they considered linking with practice a matter of some applied mathematicians, but didn't seriously think about the purpose of their own research work. (…) if the link with practice

can be indirect, one can be comfortable with one's own work even if it is removed from practice; and if mathematics is an organic whole, whose parts cannot be favoured or neglected, then what one was familiar with can be pursued as before whether it is important or not.

(Hua Loo-Keng, *et al.* 1960: 6–7)

Even in early 1958, some mathematicians still argued in writing against an overly narrow interpretation of linking theory with practice. The journal *Shuxue tongbao* asked prominent mathematicians to talk about the lessons learned from the thought reform launched as part of the Anti-Rightist Campaign (Jiang Zehan, *et al.* 1958). Hu Shihua, a mathematical logician, explained the difficulties of finding applications for all of mathematics:

Generally speaking, to keep to the socialist road in mathematics means to keep mathematics serving the socialist construction. But mathematics has its special character, its services to the socialist construction have their special form, some mathematical branches are directly connected to production, some not nearly as directly. (…) Regardless of what concrete work a mathematician does, or in which mathematical discipline his strength lies, as long as he has a clear aim to serve the socialist construction, he is following the socialist road. (…) In the United States, the thought most fashionable under the dominion of monopolistic capitalism is the most corrupt reactionary philosophy of pragmatism, which serves the needs of imperialism. It emphasizes finding empirical formulae from practice and rejects theoretical proofs, rejects the importance of mathematics, thus being a concrete manifestation of how US science follows the capitalist road. We know that scholarly thought in the Soviet Union is essentially different from this.

(Jiang Zehan, *et al.* 1958: 47)

In comparison, Hua Loo-keng's contribution to the same article was the clearest and most explicit call for theory linked with practice (in the specific forms it actually later took). This is perhaps not so surprising given that he had been required to vocally advocate it all the time, due to his prominent position, and that he needed so much to show his loyalty to socialism at this precise moment:

Mathematicians apparently need to be especially alert about the following deficiencies: 1. Not seeing the integrity of mathematics. (…) 2. Not seeing the collective character of mathematics. (…) 3. Not seeing the importance of links with practice, thinking that only highly abstract questions are worth studying. As for the problems from practice, either they look down on them as not 'beautiful' enough, or they are afraid of complications and not willing to start working on them. (…) I would also like to provide some not yet ripe suggestions: 1. Firmly establish the communist life outlook and world outlook. (...) 2. Frequently study Marxism–Leninism. (...) 3. Unite with workers and peasants and cultivate, through the exercise of labour, common emotions with them. 4. When coming upon a problem from practice, regardless of how easy or difficult it is, one should give it a try, and also actively seek problems from people working in practice.

(Jiang Zehan, *et al.* 1958: 47)

Hua Loo-Keng wanted to remedy his mistakes from the past year and reaffirm his ideological reliability. The same thing happened to Guan Zhaozhi. His self-evaluation explained:

During the Great Leap Forward in 1958, [I] moreover drew a clear line between [my]self and the bourgeois tradition, and started to think about problems of development of the natural sciences according to the needs of the Party's cause and of the socialist system.

(Guan Zhaozhi and IMCAS 1964)

Guan's 'thinking' resulted in an article that appeared in the *People's Daily* on 16 August 1958. It was called 'Mathematics also needs more of the present and less of the past' (literally 'thicker present and thinner past', *hou jin bo gu* 厚今薄古) and elaborated on a policy aimed initially at the humanities and social sciences.[14] Here, 'the past' meant non-Marxist liberal approaches and 'the present' Marxism–Leninism. Guan transformed this message to refer to old disciplines and problems on the one hand, and new problems encountered in the socialist construction on the other:

In practice, during the construction of a socialist society, if we are to leave aside unattended various practical problems in science and technology that are of a mathematical nature and concentrate instead most of our human resources and effort on the residual problems of centuries ago with little prospect of solution, then surely such actions cannot possibly be defended! It does not matter how 'delicate' such techniques are, they are detached from practical production and there cannot be much rationale in them for science and technology.

(Guan Zhaozhi 1958c: 4)

Guan's article coincided with a campaign to 'pull out white flags and stick red flags' (*ba bai qi, cha hong qi* 拔白旗，插红旗). 'White flags' were those mathematicians who actively opposed or simply did not implement the Party line of linking theory with practice. The campaign emanated from Wuhan University and spread to Peking University and CAS in early September 1958. There it expanded more broadly to the 'criticism of bourgeois academic thought'. The chief target in IMCAS was Hua Loo-Keng. Guan's article was used as a weapon by the campaigners and republished in October 1958 in the popularization journal of the Chinese Mathematical Society, *Mathematical Notices* (*Shuxue tongbao*). The same issue described the way in which the campaign moved against Hua (without naming him):

> A high tide of posters and debate arose around the question of development of mathematics. Young comrades from the Research Group of Partial Differential Equations assumed the stage first, raising two big questions for discussion: (1) Wasn't the thought guiding the development of world mathematics in the last hundred years wrong? In particular, wasn't the thought guiding some influential mathematicians of our country in the past nine years wrong? (…)

> Then we discussed the situation in mathematical circles during the nine years since Liberation, and everyone thinks that even until now, bourgeois idealist academic thought still occupies the leading position in our mathematics. Mathematicians have walked down as before the Euro-American bourgeois academic path from theory to theory, of papers for the sake of papers. When we analyzed more than 240 papers published in *Acta Mathematica Sinica* in the eight years since Liberation, none of them was linked to practice, none of them solved a problem of Chinese reality.
>
> (IMCAS 1958a: 5)

Another member of IMCAS criticized publicly in *Shuxue tongbao* was the topologist Zhang Sucheng. Zhang taught a course for philosophy students at People's University (*Zhongguo renmin daxue*) and summed up his introduction in a journal article (Zhang Sucheng 1958). One of the readers answered Zhang's formal call for comments under the pseudonym *Chengqing* (literally 'Clarification' 澄清, an aggressive-sounding pen name implying clearing of deliberately murky waters). Zhang was criticized for what he wrote about 'the things that geometers are most concerned about':

> What Mr Zhang says geometers are concerned about and pay great attention to is, I must say as a mathematician, indeed so. But at the same time, it is very incomplete, and some of the extremely important concerns they should pay great attention to are missing. Our motherland is now leaping forward, there is the Great Leap Forward in agricultural and industrial production, every day equals twenty years, these should be the concerns of every scientist leaping forward on the great socialist road. And the question to be especially concerned about is this: how to link our scientific research to politics, to production and to practice? It is extremely regrettable that Mr Zhang hasn't thought about these questions which geometers should be concerned about. I deeply felt he must be doing science for its own sake here?!
>
> (Chengqing 1958: 20)

It is not clear whether Zhang was in any way affected by this criticism. Unlike Hua Loo-Keng or Wu Wen-Tsun, he was still doing his research on homotopy groups on a sphere even during the most utilitarian times of 1959 (Chinese Academy of Sciences 1959).

Often the chief punishment for old mathematicians was the removal of their students and assistants from IMCAS. Hua Loo-Keng's student Chen Jingrun as well as Wu Wen-Tsun's student Yue Jingzhong had to leave IMCAS and worked in factories until the early 1960s (Wang Lili and Li Xiaoning 1998: 88–91; Wang Yuan 1999: 246). On the other hand, some people were promoted or had their salaries raised in late 1959 in explicit recognition of their political activities. Wu Xinmou's monthly salary was raised from 207 yuan to 241 yuan, and Guan Zhaozhi became a full research professor. As mentioned before, promotions were extremely rare in IMCAS. The political assessment of the two colleagues was rather different. Wu Xinmou was seen as too soft, whereas Guan as too stubborn:

> [Wu Xinmou] is absolutely loyal to the Party's cause, accepts tasks and adversity without complaining (…). He is able to implement and execute Party resolutions and policies, obeys the leadership, has been firm and active in all campaigns and rectifications, conscientiously analyzes his faults.
>
> [Guan Zhaozhi] actively and responsibly [works for] the Party's cause, is very serious about all his responsibilities, upholds Party guidelines and policies, strictly adheres to [the principle of] Party leadership (…). In all political movements and Party rectification movements behaved relatively actively, made honest investigations of his thought and habits. Sometimes does not sufficiently accept adversity.
>
> (IMCAS 1959c)

The painful political campaigns of 1957 and 1958 started an experiment with increased 'links to practice' in IMCAS. It did not last long and had to be rebalanced in the early 1960s. Let us now look at this story in more detail.

Leaping from theory to practice and the post-GLF adjustment

The forced practical reorientation of 1958 affected both individual researchers and the organization of IMCAS. Mathematicians first randomly collected problems from practice in various departments – this was called 'to run after tasks' (*pao renwu* 跑任务). This practice escalated in the summer of 1958, when several dozen visiting scholars joined the Institute. They were mostly fresh graduates assigned as mathematics teachers at various Chinese universities.

In October 1958, a 'General Staff' (*zong zhihui bu* 总指挥部) was formed to direct the institute, and Fan Fengqi 范凤岐 (1908–84), a veteran CPC and People's Liberation Army cadre, was appointed head of the Party group.[15]

He came with a mission to promote theory linked to practice, clashed about it with senior researchers, and eventually failed. In 1964, shortly before being removed from IMCAS, he complained about 'little achievement in the past 5 years, many mistakes' in a self-assessment report:

> Research direction of IMCAS has still not been resolved at present. In many disciplines, research still goes on by quoting and relying on classics, according to classical methods and personal interest. This is out of line with the construction of our country and mathematics thus cannot wield the power of its applications.
>
> (Fan Fengqi 1964; Fan Fengqi and IMCAS 1964)

Fan's reorganization of the institute showed his military background. He dissolved the research groups[16] and grouped researchers into four large sections. The first included mainly analysts and differential equations specialists, the second statisticians, the fourth and smallest mathematical logicians and computer scientists. The third section was oriented towards operations research (*yunchouxue* 运筹学), a newly fashionable discipline introduced by the rocket scientist Tsien Hsue-shen and other recent returnees from the USA. It was headed by a newly appointed research professor and CPC cadre Sun Keding 孙克定 (see Table 3.1) and Fan Fengqi himself. Many algebraists, number theorists and topologists, including Wu Wen-Tsun, were assigned to this section to develop operations research as a new field of Chinese mathematics.

How did researchers cope with the requirement to link theory with practice? For example, Lu Qikeng (Cheng Minde 1994: 521–37), who had studied functions of multiple complex variables under the direction of Hua Loo-Keng, turned to field theory under pressure from his Peking University students. Some of them asked him to explain the origin of the theory of functions of complex variables, and when he was only able to point to another mathematical theory, they reminded him of Mao's words that 'true science comes from practice and can direct practice'. To avoid having his subject labelled as pseudo-science, Lu turned to the theoretical physicist Zhang Zongsui, who told him that functions of complex variables were useful for the study of dispersion relations.[17] Lu learned more from Zhang's graduate student Dai Yuanben 戴元本, and could return to his class with a proof that 'dispersion relations were useful, so the functions of multiple complex variables were useful too, and not a pseudo-science!' (Lu Qikeng, interview with the author, 7 October 2008).

Lu Qikeng followed on from the work of the Soviet mathematician and theoretical physicist Nikolay Bogolyubov, director of a laboratory at the Joint Nuclear Research Institute in Dubna, who applied the edge-of-wedge theorem from the theory of complex functions to investigate dispersion relations. This direction proved to be quite fruitful for several years even after the Great Leap Forward (Lu Qikeng 2008).

Another of Hua Loo-keng's students, Wan Zhexian 万哲先 (born 1928), an algebraist (Cheng Minde 1998: 401–10), applied his algebraic expertise to operations research. During 'running after tasks', he and his colleagues discovered an empirical method for optimizing transportation distribution used by transportation planners. The method was effective but did not have an established proof (Wan Zhexian, interview with the author, 7 October 2008). The story served the requirements of the time very nicely and was retold in the introduction to the article explaining the proof in the journal *Shuxue tongbao*:

> When we contacted a department of grain this September [1958], we learned from the comrades who actually plan distribution of grain a method which they called 'operation diagram method' *tu shang zuoye fa* 图上作业法. They told us that this method has been recently widely adopted by comrades planning grain distribution throughout the country. (...) But they did not have a theoretical basis for this method; they wished that we would find one for them, one that they could understand.
>
> (IMCAS Linear Programming Group 1958a: 2)

The graphic algorithm was quite simple: it simplified the transportation network into a system of connected circles, assigned an initial feasible distribution plan to the links (such that demand in all nodes was satisfied and no link was traversed twice), and then incrementally improved the plan by balancing clockwise and counter-clockwise flows in each circle (see Figure 3.3).

Researchers did not find any proof of the method in the literature. Moreover, they realized that the problem was related to the Monge–Kantorovich Transportation Problem, solving it more efficiently than other known methods.[18] The fact that the inventor of the method was not known revealed that 'the masses are not as mean as the bourgeois experts, who always want to put their name on a method they have created.' It was emphasized that the masses can also create mathematics and implied that professional mathematicians should work humbly with them.[19]

The Transportation Problem provided other opportunities to link theory with practice. Wan Zhexian and the ARP Yue Minyi devised a proof of a second popular method used in transportation problems, the 'operation table method' (*biao shang zuoye fa* 表上作业法), which was more efficient for larger problems (IMCAS Linear Programming Group 1958b; Wang Yuan 1999:249–50}. For a few months, the Transportation Problem, linear programming and other topics from operations research (*yunchouxue* 运筹学 literally means 'study of transportation and planning') became the most attractive topics for many Chinese mathematicians.[20] One reason was that it was relatively easy to enter these subjects:

> Members of the group were mostly young comrades formerly from abstract disciplines (algebra, topology, number theory), and the majority

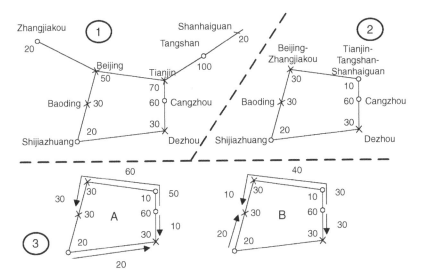

Figure 3.3 The 'Operation Diagram Method' illustrated on a schematic railway network in north China

Legend: (o) – points of supply, (x) – points of demand, numbers – volume of supply/demand in trains. The actual network (1) is simplified to cycles by reducing non-cyclic branches to additional demand or supply at the branching point (2). The bottom figure illustrates that distribution plan B, with the length of clockwise and counter-clockwise flows more or less balanced, is more efficient than the less balanced plan A. Based on IMCAS Linear Programming Group (1958a).

were young teachers sent by universities this summer for training. They were truly absolute novices to linear programming, some people stepped off the train in the morning, arrived in the institute, deposited their luggage and already in the afternoon participated on the busy work on tasks. (…) Although they were not familiar with their subject, they had courage, learned on the job with diligence and perseverance, held frequent brainstorming meetings, where those who knew how taught the others, and thus even though there were no experts, they overcame all the various difficulties by the collective mass wisdom and power.

(IMCAS 1958b: 7)

A high point of 'linking theory with practice' was the 1958 Chinese National Day, anniversary of the establishment of the People's Republic on 1 October 1949. It was a custom to 'present gifts' to the party and the people on important festivals, so IMCAS researchers pledged to 'fight hard for twenty days to produce concrete results in linking theory with practice, to be presented as a gift to the Party for the National Day'.

Among the 41 'gifts' were the computation of the national transportation plan for the fourth quarter of 1958, saving 60 million yuan compared to the plan prepared using traditional methods; flood modelling for the Three Gorges dam; a scheme to link all hydropower plants and use their output more rationally; improvement and better theoretical understanding of large-span shell roofs; and research into long-term weather-forecast models. A report on the campaign in the journal *Shuxue tongbao* highlighted the work in linear programming and differential equations ('very significant both theoretically and practically'), and emphasized the political implications:

> Many of the mathematical problems we discovered in production practice could not be found in books and documents. (…) We have profoundly experienced that practice is the only source of theory and that theses such as 'mathematics has its specific laws of development' and 'mathematics can not be linked with practice' are bourgeois, anti-scientific and reactionary positions.
>
> (IMCAS 1958b: 8)

One achievement of the movement eventually entered operations research textbooks. In 1958, Guan Meigu (管梅谷, born 1934), a young university teacher in Shandong Province, tried to mathematically optimize mail deliveries, and formulated the problem in terms of graph theory as finding the minimum-cost trail passing through all edges of a graph at least once. He used a version of the operation diagram method to solve the problem (Guan Meigu 1960). Guan's results were published in English in 1962 and picked up ten years later by Alan J. Goldman and Jack Edmonds from the US National Institute of Standards and Technology. Goldman called Guan's problem the 'Chinese Postman Problem' for its exotic and mysterious sound (Black 1999).

Wu Wen-Tsun was also assigned to the third section of IMCAS and tried to work on modern applied mathematics. He initially studied queue theory in order to solve problems in operator-connected telephony. IMCAS work reports mention these activities:

> We familiarized ourselves this year with some actual conditions of telephony and water reservoir operations for queue theory, getting ready to accept tasks in this area and launch queue-theory research.
>
> (Chinese Academy of Sciences 1959)

But this effort produced no results, and in 1959 Wu started to explore game theory instead. His first article (Wu Wen-Tsun 1959c) gave a short proof of the fundamental theorem of game theory, which states that every finite, zero-sum, two-person game has strategies that are optimal for both players (von Neumann 1928). This theorem is of the minimax type, asserting that points where a value function over the domain of possible strategies is minimal for the opponent and maximal for oneself are identical regardless of the order of minimalization and maximalization ($\min_x\max_y f(x,y) = \max_y\min_x f(x,y)$).

Table 3.4 Citations of Wu Wen-Tsun's works on game theory

Article	MathSciNet	CNKI	WoS	Google Scholar
Wu Wen-Tsun (1959b)	14	NA	NA	54
Wu Wen-Tsun (1961a)	1	1	NA	5
Wu Wen-Tsun (1961b)	2	0	NA	12
Wu Wen-Tsun and Jiang Jiahe (1962)	25	15	58	145

Wu's proof used the topological properties of continuity and connectedness to prove a generalized version of any minimax theorem of a similar type.

Although his research moved to an area presumably closer to applications, this initial work was a fine exercise in point-set topology of theoretical, rather than immediate practical, interest. Wu's results were noticed by Vorob'ev (1970) and frequently mentioned in the theory of minimax theorems (Geraghty and Lin 1984; Kindler 1993).

After his entry into game theory, Wu (1961a) investigated equilibria of non-cooperative games, a concept introduced by Nash (1951) and Glicksberg (1952). He expanded on their work by considering restricted domains of strategies, which posed additional theoretical challenges, but also brought the theory closer to actual applied problems (always restricted in various ways). Wu used a deep topological theorem of Leray (1945), on which he also wrote another comment (Wu Wen-Tsun 1961b). N. Vorob'ev included a Russian translation of Wu's work on games with restricted domains in his edited volume (U Ven'-tszyun 1963).

Wu's most influential contribution to game theory was the concept of 'essential equilibrium' of games, introduced in the collaborative article Wu Wen-Tsun and Jiang Jiahe (1962), written together with an IMCAS research trainee. This collaborative paper is one of Wu's most cited publications in all three non-Chinese citation databases (see Table 3.4).[21]

The 'state of mobilization' was a major disruption of Wu's career. At the same time, we can see his turn to game theory as more than a response to political pressure. In line with Suttmeier's model of intermediate goals achieved in consecutive states of regularization and mobilization, Wu achieved a goal in the new field which was not possible in topology – research of social relevance. This goal was maybe partly enforced at the time but would become completely internalized in the next decade and a half.

The problems of the Great Leap Forward strategy were noticed in many areas in late 1958. The IMCAS plan for 1959 recognized the following issues:

1   Insufficient depth of understanding of the practical problems being solved hindered the discovery of new truly significant theoretical impulses.

2   Because of the urgency to finish the task, people preferred old methods or empirically established techniques over developing new mathematical theory.

3   Older and more experienced mathematicians were not sufficiently active, sometimes passively pushed around instead of learning to link theory to practice.

(IMCAS 1959a).

The planned more balanced approach for 1959 revealed specific problems of the preceding year. Guan Zhaozhi's call for 'thick present, thin past' discussed above was countered by the injunction to 'learn all good results and effective methods from China and abroad, from old times as well as modern ones'. The requirement that 'tasks drive disciplines' was interpreted more loosely, to allow for the indirect motivation from practical tasks to abstract mathematical disciplines. The tasks were also supposed to be selected more judiciously:

> We in our institute should select the major tasks at the research frontline, or such that contain a clear mathematical problem. When we are working on the tasks, we have to swiftly and economically complete them, but also cannot neglect the discovery of new mathematical theories. We have to know how to concentrate on [a task] but also how to come out [and derive abstract principles]. Even though a task has been completed, the abstraction still has to go on (…).
>
> Linking theory with practice must also be seen in both aspects of theory and practice. It does not mean that there should be no theory, on the contrary there must be [more and better] theory.

(IMCAS 1959a)

But this careful revisionism only lasted until July 1959, when Mao Zedong launched an attack on rightist opportunism at the Lushan Plenum (Lieberthal 1987: 311–22). Although the frantic search for 'tasks' was already over, active rebalancing of task-driven and discipline-based research did not materialize. The IMCAS work report for 1959 mentioned approvingly Zhang Sucheng's highly abstract results in the study of homotopy groups, but avoided the question of correct research orientation (IMCAS 1960a).

The Great Leap Forward policies created a major economic crisis in China. There was famine in the countryside but cities were affected too. Many journals stopped publication because of paper shortages, e.g. *Shuxue jinzhan* (Yang Le and Li Zhong 1996: 38–9). In 1961, the CPC announced a policy of 'adjustment, consolidation, replenishment, and raising standards' (*tiaozheng, gonggu, chongshi, tigao* 调整，巩固，充实，提高), implicitly opposed to the GLF slogan of 'more, faster, better, and more economically' (*duo kuai hao sheng* 多快好省). Although the GLF policies were never openly acknowledged as mistaken until the 1980s (Lieberthal 1987: 322–5), the emboldened critics of the experiment advocated a return to a more flexible definition of linking with practice. Specifically, the CPC Central Committee approved the so-called 'Fourteen Articles on Science', drafted by the Party groups in CAS and in the National Science and Technology Commission (CPC Central Committee [1961] 1997). This substantial document can be summarized as follows:

1 The basic task of research institutions is to produce scientific results and train scientific personnel, and their work should therefore be evaluated primarily on these rather than political grounds.
2 Research environments should be relatively stable, and each institution should make 'Five determinations' (*wu ding* 五定) of orientation, tasks, staff, equipment and organization.
3 The problem of linking theory with practice has to be approached with flexibility and foresight, making space for research topics coming from the internal development of disciplines rather than immediate practice. The CAS should focus on 'fundamental research' or those applied topics which require the highest expertise and are particularly important.
4 Plans should be prepared with due regard for the abilities and interests of the scientists, should combine short-term and long-term considerations, and should allow 'small freedoms' within the 'big plan' (i.e. some free choice of research topics).
5 Initiative should be encouraged, academic discussions and publications promoted 'to some extent', and academic standards of scientific rigour should be observed.
6 Five days of six should be spent on research rather than politics, although this can be counted as an average over a longer time period.
7 Scientific cadres should be continuously trained in their discipline, should have the necessary conditions to improve their expertise, and the best among them should be supported and rewarded more than the average.
8 Cooperation and communication between research institutions should be increased.
9 Research should be done more efficiently, purchase of new equipment avoided by using local resources, and institutions should downsize (*jiingjian* 精简) rather than continue the extensive growth of the previous three years.
10 The 'Hundred Flowers' policy should be reaffirmed, the Party should moderate discussion but avoid identifying with one side of a scientific controversy, and keep the boundaries clear between politics and scientific views, which should not be labelled as 'proletarian' or 'bourgeois', etc.
11 The Party should unite rather than divide intellectuals, and gradually educate them in 'red' values, rather than criticize them for being 'white experts'. Even the most hard to 'educate' should be supported as long as they are patriotic and serious in their scientific work.
12 More political education is needed, but it should avoid crude methods and keep scientists engaged and self-respecting.

13 Administrative cadres have to investigate thoroughly the conditions of their institutions to suggest improvements and successfully implement policies.

14 The Party hierarchy should be more disciplined and focus on political education and the most important decision-making, rather than day-to-day running of the institutions.

Most of the 'Fourteen Articles' were focused on improving research conditions and stimulating experts' initiative, but they also reflected the ongoing economic crisis in China (point 9). Central institutions in Beijing were assigned strict quotas for their staff (for which the government had to procure food from the countryside). Many recent employees were transferred to other work units or sent back to the countryside to alleviate the burden of feeding townsfolk.

IMCAS had more than 210 employees in 1959, but was supposed to downsize (*jingjian* 精简) to 140 (IMCAS 1961). Although the staff increased by 18 people between 1960 and 1962 (see Table 3.2), large numbers of short-term visiting scholars were sent away. Hua Loo-Keng used this opportunity to eliminate some of the least adept members of IMCAS. He arranged two unexpected written examinations in May 1962 for all young research trainees, and those with the worst scores had to leave the institute and sometimes even Beijing (Wang Yuan 1999: 272–3). Testing was explicitly recommended by the 'Fourteen Articles', but clearly also became an opportunity to settle political and other scores. Among those dismissed from IMCAS were Guan Zhaozhi's wife Liu Cui'e 刘翠娥 and a previously celebrated political activist Huang Zuliang 黄祖良, who had been promoted in 1959 (IMCAS 1959b). Many women were chosen for downsizing to 'look after their families' (IMCAS 1963).[22]

Because there was a shortage of mathematicians able to work in abstract subjects, many of those who had experimented with new more 'practical' topics during the Great Leap Forward went back to their former disciplines. Wan Zhexian returned to algebra, and Wu Wen-Tsun to algebraic topology, proceeding later to algebraic geometry.

Although this might be seen as the zigzagging movement of Suttmeier's model, coming back basically meant starting again, as the three lost years changed so much in abstract mathematics. Nor did power relations oscillate: the Party and its apparatus held their gains. Until 1966, IMCAS was headed by a group of CPC-affiliated mathematicians and Party officials, which did not include Hua Loo-Keng. Hua's application for CPC membership was repeatedly turned down, and the ambitious man avoided IMCAS (which he still nominally directed) and devoted his time to the University of Science and Technology of China (USTC).

Apart from the personal ambition problem, Hua also clashed increasingly with Guan Zhaozhi. The debate about the correct way to link theory with practice did not stop with the end of the Great Leap Forward; on the contrary, it was perpetuated by a lengthy exercise called the 'Three Determinations' (*San ding* 三定) – determination of direction, plan and staff, envisaged in the 'Fourteen Points'. The plans were drafted and discussed for more than two years, but in the end both Hua and Guan, who had become Deputy Director of the Institute (Chinese Academy of Sciences 1961), shunned the meetings to avoid arguments. In 1964, Guan Zhaozhi expressed his frustration that 'the progress made in the last 15 years was too small'. He confessed in his self-evaluation that he was so impatient that 'he often calculated the number of days of the next decade' and was irritable with his colleagues (Guan Zhaozhi and IMCAS 1964).

Wu Wen-Tsun did not engage prominently in the conflicts about research orientation in IMCAS, but he did advocate concentrated research into algebraic topology in 1961 (Xiong Jincheng 2010). He pleaded for stable and intensive research efforts at special meetings of the Chinese Mathematical Society ('Dragon Hall meetings' in the Beijing Summer Palace), no doubt haunted by the fresh experience of being overtaken by Haefliger in singularities research (see Chapter 2).

He responded to recent developments relevant for the theory of immersion, and published his results as a book in English (Wu Wen-Tsun 1965a). A group of young research trainees under his direction (Yue Jingzhong 岳景中, Li Peixin 李培信, Yu Yanlin 虞言林 and Xiong Jincheng 熊金城) organized teaching at the university newly affiliated to CAS (Yu Yanlin 2010).

The University of Science and Technology of China and Wu's further work in pure mathematics

The rapid expansion of China's research sector in the wake of the CPC's 'march on science' policy created shortages of talent. In the 1950s, CAS was dependent on the work assignment decisions of universities and the Ministry of Higher Education. The universities used their decision-making power to their own advantage and retained the best graduates, making only the 'second best' available to other institutions, including CAS (Zhang Zhihui, *et al.* 2008: 100).

To solve this problem, CAS established the University of Science and Technology of China in 1958, inspired by the Siberian Branch of the USSR Academy of Sciences' intention to establish Novosibirsk State University as its recruitment base (Wang Yuan 1999: 235–9). USTC was controlled jointly by CAS and the Ministry of Higher Education, and offered undergraduate courses taught by famous CAS scholars. Mathematical education was conducted in a novel system of 'a single dragon' (*yi tiao long* 一条龙) courses. Three senior IMCAS researchers presented comprehensive three-year accounts of higher mathematics to consecutive classes of USTC mathematics majors: Hua Loo-Keng (*Hua long* or 'Dragon Hua', class of 1958), Guan Zhaozhi (Dragon Guan, class of 1959) and Wu Wen-Tsun (Dragon Wu, class of 1960). Each 'dragon' was fully controlled by the teacher, who set his own curriculum and structure of the course. For example, each of the three 'dragons' used a different approach to the construction of real numbers as the theoretical foundation of calculus. Younger CAS researchers and full-time USTC lecturers led seminars and stepped in for lectures when necessary. The system of 'dragons' was repeated less strictly after 1960, and students later rarely had the opportunity to see the famous mathematicians whose textbooks they were using (Shi Jihuai 2008: 1–6, 117–9, 165–79; Luo Haipeng 2005).

Wu recalls that teaching duties were assigned to him suddenly in 1960 (Hu Zuoxuan and Shi He 2002: 69–72; Zhang Zhihui, *et al.* 2008: 99–100). Wu's course was reportedly the most difficult of the three 'dragons', reflected in the choice to introduce real numbers as Dedekind sections (Li Wenlin 2008; Shi Jihuai 2008: 4). His success can be seen in the numerous names of his students who later joined IMCAS (Li Banghe 李邦和, Li Wenlin 李文林, Wang Qiming 王启明, Peng Jiagui 彭家贵 and others).

To the students who joined Wu during this period, he was already a towering figure. Many of them had read about his award in 1957 while still at school, and most started their research and chose their direction under his guidance. They were hardly aware of his earlier French work, however. His reluctance to talk about his foreign experience was probably an attempt not to stand out, caused by the egalitarian, anti-elitist atmosphere widespread after the Great Leap Forward, and his experience of criticism in 1958 for his French visit. Only once, according to his students' recollections, did he overcome this reluctance and brought up memories of his encounters with famous Western mathematicians. These stories, intended as an exposition of the history and current research frontiers in topology, pacified a minor rebellion in his class, as many of his students complained of the uselessness of topology and demanded transfer to a less abstract subject (Xiong Jincheng 2010).

After 1961, USTC established specializations for its students in the fourth and fifth years. When Wu's class of 1960 reached the fourth year in 1963, Wu Wen-Tsun offered geometry and topology as one of four possible specializations (Shi Jihuai 2008: 5). The specialization included three courses – algebraic topology, differential topology and algebraic geometry. Wu handed over the teaching of algebraic topology to Yue Jingzhong and taught algebraic geometry instead, a subject that was less familiar for him but gave him some new research impetus.

Wu used an older textbook (van der Waerden 1939), chosen for its relatively concrete approach with a lot of examples and calculations. Van der Waerden built the concept of algebraic varieties, the central object of algebraic geometry, on the notion of *generic points* and their *specializations* (van der Waerden 1939: 110). Van der Waerden's focus made it possible to avoid topological methods and related considerations of continuity (Dieudonné 1985: 70). Van der Waerden's approach was generally considered outdated in the 1960s, supplanted by the 'seventh epoch' of algebraic geometry (Dieudonné 1985), which has since the 1950s used concepts from abstract algebra and algebraic topology such as sheaves, spectral sequences and schemes. Although the 'seventh epoch' was to a large extent started by Wu's erstwhile French colleague J.-P. Serre and the somewhat younger A. Grothendieck (Dieudonné 1985: 94–113), Wu preferred the older theory. When two of his assistants translated van der Waerden's book into Chinese, he wrote a letter to the publisher praising van der Waerden's work:[23]

> Grothendick's system of algebraic geometry centred on the concept of scheme, developed in recent years, shone brightly for a time, but lately a famous Harvard professor said that even though Harvard is a world centre of algebraic geometry, there has never been a course there on the theory of schemes. (…) Among the various schools [of algebraic geometry], van der Waerden's system around the central concept of generic points is not only the first theory that laid the foundations of algebraic geometry, but in my view is also the school that has the greatest future and has best performed in the test of history.
>
> (Wu Wen-Tsun 2008: i)

Wu's student Li Banghe (born 1942), now CAS academician and leading Chinese algebraic topologist, also defended this choice as prudent. According to Li, only very few who studied algebraic geometry directly from Grothendieck's abstract system could reach any useful results. The theory has to be supported by concrete examples, such as curves and surfaces, amply used in van der Waerden's book, which give the student an intuitive grasp of algebraic varieties. Li Banghe's explanation of Wu Wen-Tsun's motives shows that pragmatic epistemology became deeply entrenched among Chinese mathematicians and remains so even today:

> Some people think that you have to learn [the abstract theory]. My and Mr Wu's view is not quite similar to those people. If you want to do research, you have to understand concrete examples. And theory has to be abstracted from concrete examples.
>
> (Li Banghe, interview with the author, 8 July 2010)

Wu's conservative approach was also reflected in the methodology of his two substantial papers on algebraic geometry (Wu Wen-Tsun 1965b, 1965c), which were based on Hodge and Pedoe (1947–54), an English textbook written with the same basic style, language and symbolism as van der Waerden's classic. Wu's first article proposed a new construction of Chern classes for algebraic varieties with singularities. Algebraic varieties are the counterparts of manifolds in algebraic geometry, and Chern classes are especially useful for their investigation. Wu's construction was the first that enabled the actual calculation of Chern classes and Chern characteristics for algebraic varieties with arbitrary singularities. Although noticed and reviewed by S.T. Hu, it did not attract any citations and was later superseded by more general mainstream treatments, such as MacPherson (1974). On the other hand, Wu and some of his students returned to the study of algebraic varieties after 1985 within the framework of constructive mathematics.

Wu Wen-Tsun took up the theory of singularities of algebraic varieties to follow in the footsteps of two great Chinese mathematicians he had personally met: his teacher S.S. Chern and another algebraic geometer from the same generation, W.L. Chow, to whom Wu had been introduced in Shanghai by his friend Zhao Mengyang (see Chapter 2). Wu used theoretical tools from Chow Wei-Liang (1956) and insisted on the renaming of Cayley forms and Cayley points, as used in Hodge and Pedoe (1947–54), to Chow forms and Chow points.[24] Wu demonstrated an inclination to set the record straight for his nation and its mathematics, which would manifest itself strongly when he began studying ancient Chinese mathematics.

After the Cultural Revolution, Wu published another article on singularities from a topological perspective (Wu Wen-Tsun 1974). His results, already achieved in 1966, had in the meantime been replicated and developed elsewhere, albeit by different methods. Overall, Wu's work in algebraic geometry equipped him for his later study of the mechanization of mathematics, but did not generate any noticeable interest in the mathematical community. His further plans to study cohomology relations between real and imaginary parts of complex manifolds (IMCAS 1965, 1966d) were interrupted by politics again – the start of the Cultural Revolution.

Conclusions

Between 1949 and 1966, the Institute of Mathematics evolved from a small research outlet, whose members were all born between 1910 and 1930 and shared very similar backgrounds and commitment to mathematics, to an organization with more than 200 employees of various ranks of seniority, controlled by a more politicized leadership that was split over its correct future course. The 1960s saw a consensus enforced by a coalition of moderate forces within the Party to protect the status quo created by the Anti-Rightist Campaign and the Great Leap Forward. There was no suggestion of going back in terms of policy or power sharing. Although the 'Fourteen Articles' called for more democracy and freedom of research, the golden years of the 1950s never returned.

Suttmeier's description of the policy instability as conscious 'social experimenting' has problematic relevance for the Institute. Although the sources tend to portray the changes introduced during the Great Leap Forward as final solutions to long-standing unresolved problems of research orientation, on closer inspection their motivation was more or less external, a combination of the frenzied political climate with the prodding of activist figures like Guan Zhaozhi by the local party leadership. Major changes in IMCAS were hardly ever motivated by urgent problems within the institute. This tension is only stronger on the level of individual careers, as can be seen on Wu Wen-Tsun's case. The policy changes always interrupted his research at promising moments and forced him to abandon projects before they could bear fruit.

Although the exact decision-making process and the higher administrative levels' degree of involvement cannot yet be reconstructed, the IMCAS mathematicians' massive reorientation to tasks of immediate practical relevance in this period appears to be more a side effect of the GLF than a centrally planned experiment. Of course, individuals had to experiment in the face of this pressure, and the preference for certain disciplines (game theory, operations research) can be seen as an experiment on a larger scale. But more successful conscious experimentation can perhaps actually be found in the quieter 'regularized' periods in the mid-1950s and during the Readjustment after 1961.

In 1958, the Party managed to evoke an atmosphere similar to total war. This was familiar territory for its leadership, hardened by decades of guerrilla warfare. Overcoming bureaucratic and other obstacles in 'academic thought', as well as current technological and economic constraints, was seen as a fight requiring discipline and the total commitment of every individual. What was merely a convenient metaphor or a conceptual framework for the CPC leadership, however, was a direct challenge to those CPC members who, like Guan Zhaozhi, had not taken part in the actual pre-1949 fights. They had to prove their discipline and commitment this time.

In power terms, the Party let one senior mathematician (Hua Loo-Keng) and his ideas dominate the institute in the 1950s, and relied heavily on another one (Guan) during the Great Leap Forward, but afterwards the CAS's CPC committee tried to limit the influence of both men rather than giving Hua the upper hand again. The leftist pressure inside and outside the Institute was permanently strong and influenced both research orientation and debates.

Recent scholarship on mathematics in the Soviet Union can be used to draw a contrasting comparison to China.[25] After Stalin assumed power in 1928, there was also a period of ideologically loaded attacks on pure mathematics in the name of dialectical materialism, with arguments and rhetoric similar to the Chinese writings of the 1950s – quotations from Lenin's *Materialism and Empiriocriticism*, references to class struggle in mathematics and to the proletarian line. The USSR, however, had not only many more world-class mathematicians, but also more sophisticated ideologues interested in the philosophy of mathematics ranking higher in the Party hierarchy than Guan Zhaozhi. The most prominent example was Ernest Kolman (1892–1979), an Austrian-born émigré with unfinished degrees in philosophy and engineering, and in the 1930s a 'Red Professor' at the Moscow Communist Academy. Kolman's attacks were not against mathematical research per se, but against the religious and philosophical affiliations of mathematicians such as D.F. Egorov (1869–1931) and Nikolai N. Luzin (1882–1950). Unlike in China, younger equally talented mathematicians – rather then merely politically qualified upstarts – used these campaigns to assume positions of leadership. An episode more reminiscent of China occurred in Leningrad, where a group of activists ('mathematicians-materialists') led by a teacher of philosophy L.A. Leifert attacked the local mathematicians, including authoritative Party members such as O.Yu. Schmidt. They managed to completely destroy the Leningrad Mathematical Organization in 1931, which was then only re-established in 1959, but their success was short-lived, as Leifert had already, in 1932, been criticized by Kolman's Communist Academy for superficiality and 'leftist deviations'.

The outcome was thus quite different than in China, probably for two main reasons. In the Soviet Union, mathematicians were allowed to handle the administration of their institutes, and their professional views were thus less alien to the Party than in China. Moreover, Stalin did not rely on successive mass campaigns to secure power, and thus did not generate the cycles of mobilization and regularization pervading all spheres of life. The Great Leap Forward, on the contrary, only temporarily weakened Mao and the leftists, who continued to oppose the Readjustment policies and effectively mobilized ideology as a weapon. 'Linking theory with practice' and political control over mathematicians thus remained on the agenda throughout the 1960s and 1970s, although it was overshadowed by new and more pressing concerns after 1966.

Notes

1 On the political dimensions of the Great Leap Forward and the disaster it has created in the countryside, see MacFarquhar and Fairbank (1987), and especially Lieberthal (1987), Pepper (1987) and Lardy (1987). A more recent treatment of the famine is Dikötter (2010).

2 The *locus classicus* of the policy oscillations theory is Skinner and Winckler (1969); cf. its criticism by Nathan (1976). A similar paradigm of cycles of 'relaxation and tightening' (*fang/shou* 放收) has, however, also been used to describe PRC's post-CR development (Baum 1994: 5–9; Dittmer 1990).

3 Approaches similar to Suttmeier's are subject to 'a long list of rather standard and by now tired objections' (Oksenberg 1975). The impression of a grand plan pursued by the Chinese leadership is to a large extent due to Suttmeier's reliance on officially published policy documents, with no access to evidence about actual policy making and implementation. This led him to ignore the competing interests of different political groups and to abuse metaphors from control theory and cybernetics (Stavis 1975). In later publications, Suttmeier explained the policy changes by a lack of leadership consensus rather than social experimentation – see Suttmeier (1980: 1).

4 Suttmeier originally extended his thesis to the changes during the Cultural Revolution, which seems too evidently inappropriate in our case – the Cultural Revolution was a complicated set of almost purely political movements, and did not have any obvious goals in scientific research. See Chapter 4.

5 See especially Schmalzer (2008), a study of China's paleoanthropology at the intersection of elite and popular science.

6 (Zhang Dianzhou 1999: 199–204). Several short histories of IMCAS have been officially compiled in China, e.g. Wang Yangzong and Cao Xiaoye (2010: 76–90). The most detailed (if not always clearly written) version is, however, still Wang Yuan (1999).

7 Hua Loo-Keng's book *Additive Prime Number Theory* was allegedly considered for the Stalin Prize in 1953 (Wang Yuan 1999: 175).

8 Tsien Hsue-shen also made a humble speech upon receiving the prize. Wu Wen-Tsun, if he spoke at all, was not quoted in the *People's Daily*.

9 Zhang Jingfu and Du Runsheng have been interviewed by Liu Zhenkun (Liu Zhenkun 1999b, 1999a).

10 The story is told in Xu Lizhi, *et al.* (2009: 130–1), adding that Guan Zhaozhi organized around a hundred criticism meetings of Sun. Cf. Sun Yifeng (2010).

11 On the rectification campaign and the launch of the Anti-Rightist Campaign, see Goldman (1987a). For the scientists' perspective, see Wang Yuan (1999: 220–5), Cao (1999) and Chen Jianxin, *et al.* (1994). For a selection of self-criticisms of scientists published in newspapers at that time, see Chen (1961).

12 The campaign was fiercest at universities due to their complicated social structure (sharp divides between old and young teachers, administrators and students). See Andreas (2009: 32–41) for an account of Tsinghua University.

13 These included two collaborators of Wu Wen-Tsun, Li Peixin and Jiang Jiahe, as well as Lin Qun, now a CAS academician (IMCAS 1958a).

14 The slogan was popularized by the chief party ideologue Chen Boda (Chen Boda [1958] 1971) and by the President of CAS Guo Moruo (Guo Moruo 1958). For more on this campaign in its initial field – history – see Feuerwerker (1961: 346–8) and Weigelin-Schwiedrzik (1996).

15 Fan Fengqi had been a Brigadier (*daxiao* 大校) of the People's Liberation Army, county Party Secretary in the Communist-controlled areas in the 1940s, and volunteered to fight in the Korean War; see his obituary in Xinhua News Agency (1984). Apart from Fan Fengqi, the General Staff also included Guan Zhaozhi, Wu Xinmou and Hu Shihua, the new arrival Sun Keding, the head of the IMCAS General Office Zheng Zhifu 郑之辅, all CPC members, as well as Hua Loo-Keng and a young activist Zhang Zhaozhi 章照止 (born 1933). In 1959, the General Staff was replaced with a larger group including 2–3 people from each research section, including non-CPC senior mathematicians such as Zhang Zongsui and Hiong King-Lai. (Hua Loo-Keng and Fan Fengqi 1959).

16 Ordinary Differential Equations, Partial Differential Equations, Functional Analysis, Probability and Statistics, Mathematical Logic, Algebra, Number Theory, Function Theory, Geometry and Topology, Mathematical Physics, Applied Mathematics (IMCAS 1958d).

17 Dispersion relations describe the propagation of wave–particles through space depending on their frequency. Refraction and absorption properties of different media can be described by a single complex dispersion index, which makes functions of complex variables relevant.

18 There are several possible methods of solution of the transportation problem. One very similar to the Chinese operation diagram method had been published in the Soviet Union by A.N. Tolstoi in 1930. The method most widely used now, published by Ford and Fulkerson in 1955, is similar to the other method created during the GLF, the 'operation table method'. Both Tolstoi and Ford–Fulkerson methods were in publications that were probably inaccessible to the Chinese mathematicians in 1958 (Schrijver 2002).

19 All ideological content was later removed from the introduction in the collection of Wan Zhexian's popularization articles (Wan Zhexian 1997: 151–63).

20 Another group from the Beijing Normal University published a follow-up article on the 'operation table method', mentioning their practical experience from a month-long internship in the Yuquan Road Truck Depot in Beijing (Beijing Normal Institute 1959).

21 Wu also wrote a popularizing article (Wu Wen-Tsun 1959a), planned to form the first part of a series, which did not continue. This article contained an allusion to forerunners of game theory in Chinese historical literature – see Chapter 6.

22 In one of the waves of downsizing, ten employees were made redundant, of whom seven were women (IMCAS 1962b). Most were clerks, but two had worked as research trainees in 1960 (there were 10 women among 141 researchers). Wu Wen-Tsun was also hit by downsizing as his assistant Sheng Jiting 盛继庭 had to leave the institute.

23 The letter is reprinted in Wu's preface to the 2008 Chinese edition of van der Waerden's book. It is not clear when it was written. It mentions the 'recent reprinting' of van der Waerden's *Introduction to Algebraic Geometry*, which happened in 1973. Wu's letter supported its Chinese publication which was eventually cancelled – that would suggest the politically turbulent 1970s.

24 This has since become the common name, as in Dieudonné (1985).

25 See Lorentz (2002) and Seneta (2004) for accounts of the campaigns in the 1930s. See also the detailed study of the persecution of the religious 'School of Names' (Graham and Kantor 2009).

4   The transformative effect of
the Cultural Revolution

In the November 1976 issue of the newly established *Journal of Applied Mathematics* (*Yingyong shuxue xuebao*), Wu Wen-Tsun, one of its editors, published an article entitled 'The Cultural Revolution opened up a broad future for mathematical research'. It was Wu's celebration of his intellectual transformation since the start of the Cultural Revolution (CR) in 1966, and his manifesto for the future. After invoking the current politically correct slogans and briefly summarizing China's achievements in applied mathematical disciplines, Wu turned to his personal experience. In order to highlight the depth of his transformation, he started with a thorough self-criticism:

> I am an intellectual coming from the old society, I studied abroad, and I carried over to the new society the modes of study and research I had learnt in the old society and abroad. In the past, I always aimed my effort at creating some new theory, establishing some new school, and my mode of research exhibited the three separations [from politics, from the masses and from practice] and was academic.
>
> (Wu Wen-Tsun 1976: 14)

The Cultural Revolution offered Wu three great opportunities: to study in detail Mao Zedong's and other Marxist–Leninist works, to engage in factory labour for a few months, and to discover and understand the greatness of ancient Chinese mathematics. Wu argued that his research became much more socially relevant due to these three transformative experiences, and promised to keep drawing on them in the future:

> Of course, this work has just begun and still continues. But we firmly believe that under the guidance of Marxism–Leninism and Mao Zedong Thought, with unceasing effort to learn from the production practice of the labouring masses, and while absorbing positive inspiration and nourishment from ancient Chinese mathematics, we will be able to expand the gains of our struggle.
>
> (Wu Wen-Tsun 1976: 16)

The article appeared one month after the arrest of the 'Gang of Four' (Jiang Qing, Zhang Chunqiao, Yao Wenyuan and Wang Hongwen) on 10 October 1976, which is now officially regarded by the CPC as the end of the CR. Wu accordingly inserted – possibly at the last minute – some harsh comments about the damage the Gang had done to Chinese science by using ideology to 'suppress research workers'. But otherwise this article was an affirmation of the last decade. The leftist radicals were weakened but still dominant at lower levels of leadership, and the new CPC Chairman Hua Guofeng was trying to found his authority upon reverence for Mao Zedong and his legacy. But there was more to Wu Wen-Tsun's article than political expediency or necessity: even in his later writings and pronouncements, he continued to emphasize the transformation he had undergone, and the importance of *all three* factors which he first laid out in this article.[1]

Wu Wen-Tsun's search for a long-term research orientation, which had started in the 1950s, was for a long time unsuccessful, and continued throughout the CR period. Only at its very end did he finally venture upon a productive field and become able to effectively claim its significance. This chapter will focus on the way towards this result, but will also try to capture the complexity of the social and political environment in this period through broader descriptions of the atmosphere at the Institute of Mathematics.

Writing about the Cultural Revolution – methodological questions

The term 'Cultural Revolution' is commonly used in both a broader and a narrower sense. In the narrower sense, only the tumultuous events of 1966–9 constituted the revolution, followed by a period when the CR was viewed as a successfully finished transformation that had created a new socialist society. This was the official interpretation of the Ninth CPC Congress of April 1969, authoritative until at least the very end of 1978.[2] It is also the favoured definition of those historians who try to sharply separate the mass movement of 1966–8 from the centrally orchestrated repression and political campaigns characteristic of the 1970s. It accords with the common-sense notion of a revolution as a rapid social and political change involving illegal power seizure.

However, mainstream histories generally accept the broader meaning as defined by the CPC Central Committee's *Resolution on Certain Questions in the History of Our Party since the Founding of the People's Republic of China*, which says that 'the "Cultural Revolution" lasted from May 1966 to October 1976', i.e. from the 'May 16 Circular', which officially announced its launch, to the arrest of the Gang of Four (10 October 1976). I will use the term 'the Cultural Revolution period' in an even broader sense, to include the period up to the final victory of Deng Xiaoping and his pragmatic policies in December 1978.[3] This incidentally covers the time when Wu Wen-Tsun reached and published his first results on mechanical theorem-proving (1977–8), which will be discussed separately in Chapter 6. The 'CR period' in this conception includes two periods, a revolutionary and a more-or-less orderly one, almost in line with Suttmeier's model of consecutive states of mobilization and regularization.

Sources for the history of the CR period have become increasingly available to researchers in recent years.[4] Most of these focus on high- and mid-level politics, the Red Guard movement and radical activism in general, and, to a much lesser extent, on the persecution campaigns and their victims. The impact of the CR on scientists and scientific institutions has been described in a range of articles and biographies, and most recently also in English in the collected volume by Wei and Brock (2013).

Official histories of Chinese science and of CAS, such as (Chen Jianxin, *et al.* 1994) or (Fan Hongye 2000), are as a rule quite general, in line with Deng Xiaoping's recommendation that is it 'appropriate to be general and not appropriate to be specific' (*yi cu bu yi xi* 宜粗不宜细) when discussing the Cultural Revolution. A lot of attention is paid to the scientific results that were achieved 'despite difficult conditions'.

Wang Yuan's biography of Hua Loo-Keng (Wang Yuan 1999) fills a considerable part of this gap by chronicling the Cultural Revolution in the Institute of Mathematics of the Chinese Academy of Sciences (IMCAS) with the use of some original documents held at the Institute's 'CR Archives'. Another focus of the IMCAS history of this period is Chen Jingrun 陈景润 (1933–96), who derived and published (in 1973), despite many obstacles, a sensational result on the Goldbach conjecture. His melodramatic story has been a topic of several books, journal and newspaper articles. These offer many details about the functioning of IMCAS during the post-1971 period.[5]

These sources do not discuss Wu Wen-Tsun, whose story was less conspicuous. He was neither seriously persecuted nor criticized during the Cultural Revolution, which makes the reconstruction of the daily life of IMCAS more relevant for understanding his case. In July 2010 I had the opportunity to systematically study documents from the CR period held in the Archives of CAS (but not the politically sensitive 'CR archives' mentioned above, which have not yet been handed over to CAS Archives by the Institute). There were general summaries and plans, and reports derived from the day-to-day operation of the institute. They show both the current political climate and important details about the organization of research and other activities in the institute.

Detailed information about the the CR in CAS can be obtained from the reminiscences of Du Junfu (2008) from the Institute of Physics of CAS (IPCAS) as well as other articles from the electronic journal *Jiyi / Remembrance*. Du often compares the situation in his institute with that in IMCAS and provides details about the political struggle within the headquarters of CAS (*yuanbu* 院部).

A history written without full access to all available documents is of course bound to be partial, and at times seriously inaccurate. But I believe there is a case for writing a history of the CR in IMCAS nonetheless, as Wu Wen-Tsun's story cannot be told without an understanding, however preliminary, of this period.

Cultural Revolution in the Institute of Mathematics, 1966–69

The 'Great Proletarian Cultural Revolution' had a double dynamic. On the one hand, it was a series of campaigns initiated by Mao Zedong and other CPC leaders, with a combination of long-term and short-term goals. On the other hand, its actual conduct was largely in the hands of semi-autonomous actors (mass organizations and local civilian and military officials), often ignorant of the real goals of these campaigns, and inevitably bringing in their own interests. For most ordinary people, it was largely the latter dynamic that shaped their CR experience, which was greatly variable across the country and even across large institutions such as CAS.[6]

Little source material is available for this tumultuous period – few records were kept (or made accessible),[7] and few witnesses are willing to talk about their experiences. Almost nothing from the available evidence directly concerns Wu Wen-Tsun. I will therefore focus on IMCAS as a whole. Some of Wu's close colleagues and students were denounced or persecuted, some were radical activists, and this undoubtedly shaped Wu Wen-Tsun's attitudes to the Cultural Revolution.

In Table 4.1, a chronology of major stages of the Cultural Revolution summarizes the passage from top-level power struggles to campaigns affecting everyday lives.

The active role in these political struggles gradually shifted from activists of 'mass organizations' (Red Guards), incited by radical ideologues or establishment leaders, to those cadres and military officials who survived the initial campaign. It is now generally recognized that the renewed concentration of power after 1968 led to more violence, persecution and even mass killings, especially in some provinces (Su 2006).

Although the 'campus movement' in June 1966 started on university and college campuses, it quickly spread to other places, including the institutes of CAS. Spatial proximity as well as personal ties between institutions were crucial for initiating the movement. Graduate students and young research trainees in particular made frequent visits to their former universities, read big character posters, engaged in discussions with friends still studying there, and attempted to initiate equally exciting critical movements in their own institutes. In June 1966, there were 15 graduate students and 167 research trainees (70 new since 1963) in IMCAS (IMCAS 1966a). The young IMCAS activists had easy access to the centres of revolution at universities in the north-western district Zhongguancun, and were reportedly among the most radical in CAS (Du Junfu 2008: 6).

These leftist radicals were discontented with traditional party-led movements and intended to overthrow the 'capitalist roaders within the party'

Table 4.1 Stages and campaigns in the violent phase of the Cultural Revolution (1966–69)

Stage and Time	Description
Campus movement June 1966	University students incited by young radical officials (and ultimately by Mao Zedong) attack establishment cadres. The Red Guards form in schools and spread to universities.
Work Teams June–July 1966	CPC Central Committee dispatches work teams to take the initiative and control excesses. Work teams incite pro-establishment activists to form Red Guards ('Conservative' in view of their opponents) and attack activists of the first phase. 'Class enemies' and 'rightists' are often persecuted by the work teams.
Smashing the Four Olds August–October 1966	Mao Zedong arrives in Beijing and orders withdrawal of work teams. Rebel and conservative Red Guards form in large numbers, attack 'old ideas, old culture, old customs, old habits'. Public security is forbidden to act against the Red Guards.
Opposing the bourgeois reactionary line October–December 1966	Mao Zedong encourages attacks on ever higher levels of the Party hierarchy for previous attempts to block the Cultural Revolution. Leading cadres criticized at mass meetings. Red Guards search government offices and private homes for evidence of misconduct. Conservatives disintegrate, 'rebel' Red Guards split into factions.
Factional fighting January 1967–8	'Rebel' Red Guard factions vie for control over their institutions. In the 'January Revolution' of 1967, some factions seize power, sometimes with official sanction by higher leaders, but rival factions fight back, sometimes with arms. The People's Liberation Army (PLA) is dragged into the fighting and factionalized.
'May 16 Conspiracy' September 1967–75	Mao Zedong lets the Premier Zhou Enlai suppress some radical Red Guard factions. Security forces detain activists alleged to form a 'May 16 Conspiracy'. Investigations of alleged conspirators are often used as a tool of factional fighting and retribution.
Cleansing the class ranks 1968–9	PLA forms 'Workers and Soldiers Mao Zedong Thought Propaganda Teams' to control institutions torn by factional conflict among Red Guards. Revolutionary Committees are formed, Red Guards disbanded, and students sent to the countryside. Security apparatus targets problematic individuals, including some Red Guard leaders. Party Congress in 1969 celebrates the success of the Cultural Revolution. Cult of Mao Zedong, border skirmishes with the Soviet Union provoke war scares.

Sources: According to MacFarquhar and Schoenhals (2006) and Du Junfu (2008).

through mass rallies. They achieved quick success when the Party Secretary and Deputy Director of IMCAS, Zheng Zhifu 郑之辅, was dismissed from the Deputy Director's post on 6 June (Chinese Academy of Sciences 1966). Zheng Zhifu had come to IMCAS in 1956 and became Deputy Directory in 1962 with the task of correcting errors committed during the Great Leap Forward. During the 1964 CAS Party Committee cadre evaluation, he was both praised as someone who 'cares about intellectuals' life, pays attention to work style, and understands the special character of intellectuals', and criticized because he 'wants to save [other people's] face, is good in self-criticism but insufficient in mutual criticism' (Zheng Zhifu and IMCAS 1964),[8] which explains why he became a target of the radicals. He was harassed and 'investigated' until November 1973, with the final verdict 'serious capitalist roader' only lifted in 1978 (IMCAS 1972b, 1978).

In late June 1966, the CPC Central Committee attempted to control unfolding events by dispatching work teams to lead the Cultural Revolution. These tried to deflect the movement towards attacking traditional targets ('class enemies', 'rightists', etc.), and to crack down on the most provocative radicals. The radical 'rebels' in fact included some politically disadvantaged members, who had been persecuted before, and who were often attacked during the work-team period (Du Junfu 2008: 5).[9] One victim of the work teams in IMCAS was Xiong Jichang 熊纪长 (1941–66), a USTC graduate who had worked at the Daqing oilfield but was sent back to Beijing as a punishment for 'spreading [intellectual] poison'. He died in July 1966 after jumping (falling?) from the fourth floor of the CAS dormitory, by a later verdict 'persecuted to death by the bourgeois reactionary line' (Wang Yuan 1999: 347; IMCAS 1978).

After the withdrawal of work teams upon Mao's insistence, spontaneous 'rebel' Red Guards took their initiative from the early activists incited by the work teams. IMCAS had a 'rebel' group called the *Torch Troop* (*Huoju da dui* 火炬大队), which 'consisted of young intellectuals with exciting ideas, but there also were senior members such as Guan Zhaozhi'. Core Party activists, on the other hand, formed the *Red Guard Troop* (*Hongwei da dui* 红卫大队; Wang Yuan 1999: 296). Among the 'rebel' leaders were young USTC graduates, such as Zou Xiecheng 邹协成 (1939–70), who became a member of the CAS Standing Revolutionary Committee formed on 30 July 1967 (Fan Hongye 2000: 187).

During late summer and early autumn 1966, 'bourgeois intellectuals' came under heavy criticism and their homes were searched for implicating material. Hua Loo-Keng was dragged to a struggle session on August 20 and his former students were forced to criticize him (Wang Yuan 1999: 293–4). Wu Wen-Tsun was also declared a 'bourgeois academic authority'. Many of his students expressed wishes to leave his subject and, if possible, be transferred to military-related research institutions. Some of them emphasized the impracticality and ideological backwardness of topology in their transfer applications:

Chairman Mao teaches us that mathematics is derived from physical models, and should go back to physical models. According to the Chairman's teachings, I think I should change my current troubled position in mathematical research. I would like to go to a different unit and work on different projects. If possible, I would like to leave IMCAS. I think that I could in this way do work of some use for the people. (...) The abstract topology group has produced ten people, this is entirely a 'white expert' revisionist research path (setting up a temple for a boddhisattva).

(IMCAS 1966c)

Wu Wen-Tsun was thus regarded by his students as a self-serving expert ('boddhisattva'), who took on students ('sets up a temple') just because he was famous, rather than in accordance with current social needs. His home was searched twice, many of his books were destroyed, as well as his correspondence with his friend Zhao Mengyang, teacher S.S. Chern, and foreign mathematicians (Hu Zuoxuan and Shi He 2002: 74; Ke Linjuan 2009: 92; Wu Wen-Tsun 2006: 465).

The Red Guards at the Institute of Mathematics were formed entirely of adults, which possibly prevented the most extreme excesses reported especially from middle schools. The Red Guards' attention soon shifted from mundane problems within the institute to the politically more significant CAS headquarters, especially to the CAS party secretary Zhang Jingfu and his deputy, head of the CAS Secretariat Du Runsheng (Wang Yuan 1999: 296; Du Junfu 2008: 6).

In early 1967, 'rebels' finally 'seized power', first in the CAS headquarters and then in the institutes and research divisions. CAS headquarters were even taken twice, first on the night of 21 January 1967 by the *Red Flag Liaison* (*Hongqi lianluozhan* 红旗联络站), then on 24 January by the *Revolutionary Rebel Corps* (*Geming zaofan tuan* 革命造反团), which received approval from Zhou Enlai.[10] After about three weeks of fighting, both groups eventually formed a 'United Power Holding Committee' (Du Junfu 2008: 15).

IMCAS was controlled by the *Red Torch* faction, part of the more radical *Red Flag Liaison*. Their period in power lasted till October 1968, when the PLA-formed *Capital Workers and PLA Mao Zedong Thought Propaganda Team* (*Shoudu gongren Jiefangjun Mao Zedong sixiang xuanchuan dui*首都工人解放军毛泽东思想宣传队, 'Workers Propaganda Team' in short) occupied CAS (Wang Yuan 1999: 297).

Some research activities continued throughout 1966 and 1967, for example nuclear-related research at IPCAS (Du Junfu 2008). In August 1970, CAS research institutes were asked to report on the results they had achieved since the start of the Cultural Revolution. Guan Zhaozhi, who was then in charge of the Institute, returned the circular with a handwritten comment: 'Results not reported. There is nothing worth reporting anyway, they are relatively scattered' (Chinese Academy of Sciences 1970). In February 1971, the Revolutionary Committee submitted a more substantial reply, mostly focusing on post-1969 work, but also with several projects continuing from the pre-CR era in the areas of control theory (equipment for the military) and statistics (quality control). Their results were not implemented, however, due to the chaos within the cooperating organizations (IMCAS Revolutionary Committee 1971).

These officially solicited and organized research activities were mostly pursued by younger IMCAS researchers. Some very limited non-sanctioned research also took place. Wu Wen-Tsun's case was rather curious: during a boring political meeting held in 1966 in the IMCAS library, he picked up the latest issue of *IEEE Transactions on Circuit Theory* from the shelf and became interested in an article (Fisher and Wing 1966) about planar realizations of graphs representing integrated circuits.

Integrated circuits (IC) consist of thousands of elements connected by printed conductors. When these conductors intersect, they have to cross each other on different planes to avoid electrical contact. It is thus economically and technically desirable to limit the number of intersections to a minimum.

This problem translates readily into the language of graph theory – how can a non-planar graph be turned into a planar one with the fewest possible changes? Fisher and Wing designed an iterative method to decide whether a graph is planar or how it can be made planar. A given graph was decomposed into connected pseudo-Hamiltonian components, each of which was checked separately for edges which had to be on the opposite sides of the central circle of the component ('alternating'). If these edges could be divided into two disjoint sets, the graph was planar, otherwise non-planar.

Wu Wen-Tsun proposed a method based on the calculation of characteristic cocycles of symmetric products of all vertices and edges, as a straightforward application of his general theory of realization, immersion and imbedding of complexes for the planar case.

All information about the possibility of a planar realization could be derived from the solution of a set of equations determining the nullity of imbedding classes. This general approach was, however, unworkable in real conditions, because for a graph with m vertices and n edges, as many as n^2 equations with nm variables would have to be solved. Although the unknowns were all binary (Wu worked, as always, with modulo 2 arithmetic), it was still necessary to substantially reduce the number of equations. Wu thus introduced a series of simplifications of the original equations, based on a choice of an arbitrary oriented spanning tree of the original graph, and separation of superfluous variables. This enabled a reduction to at most $2(m - n)$ variables.

Wu's article was first written in 1967, but only published (Wu Wen-Tsun 1973) following the establishment of new journals after the Cultural Revolution. Although it demonstrated concerns for practical utility and algorithmic efficiency, it was still a theoretical work consisting of theorems and proofs. Wu, however, wrote another exposition of the same method for the Chinese edition of his book about imbedding (Wu Wen-Tsun 1978c), which went several steps further in the applied, constructive and popularizing direction. It included many more diagrams to illustrate the method, and Wu structured his exposition into 'procedures' (*shu* 术), emulating the traditional Chinese mathematical style of problem–algorithm complexes. The method was essentially manual, and still heavily reliant on a complex set of concepts and notations that would not be accessible to even a well-educated general public. But the efficacy of his method was appreciated in China – Wu Wen-Tsun (1973) has eight citations in CNKI from 1979 and later.[11]

Criticism of IMCAS senior researchers reached its height in November 1967, after the formation of revolutionary committees. Both Party members and non-Party mathematicians were criticized at mass meetings (Wang Yuan 1999: 298–300). The victims of the November 1967 meeting were put under surveillance in a 'dictatorship group' (i.e. group under the proletarian dictatorship) body of 17, including higher CAS officials. The members were required to constantly study Mao Zedong's writings, write self-criticisms and criticize each other. The younger number theorist Chen Jingrun, criticized for not paying attention to politics, was also placed into the group in April 1968. In September 1968, after a new round of humiliating harassment from the Red Guards, who even destroyed some of his research notes, Chen attempted suicide by jumping from a third-floor window, hurting his legs (Shen Shihao 1997; Wang Yuan 1999: 324). The dictatorship group was disbanded after the arrival of the Workers Propaganda Team in October 1968. The members still under investigation were placed under house arrest and paid a minimum monthly allowance of 20 yuan, at a time when the minimum salary in IMCAS was 56 yuan (IMCAS 1969).

The campaign to 'cleanse class ranks', officially launched in IMCAS on 11 July 1968, escalated after the Red Guards were disbanded by the Workers Propaganda Team in October.[12] Five researchers, including the relatively senior Wang Yuan (born 1930, Hua Loo-Keng's student) and Li Peixin (1933–?, Wu Wen-Tsun's student), were persecuted for 'three-anti speech' – anti-Party, anti-socialism and anti-Mao (Wang Yuan 1999; IMCAS 1978).

The three years 1966–9 thus appear in the available record as a time of humiliation and fear for many in IMCAS. Even those among the younger generation who had initially been thrilled by the opportunity to vent their long-suppressed frustration often later faced persecution. Many of the radicals joined the movement to transform Chinese mathematics into an activity more directly and efficiently serving the people, but looking back in 1969, they could only see wasted time and disruption. Guan Zhaozhi's comment about the scattered results in 1970 shows a lot of bitterness. Wu Wen-Tsun's students who requested transfers to other units in 1966 were all disappointed and had to remain in IMCAS. Some even continued to work on topology in the 1970s.

Wu Wen-Tsun's post-Cultural Revolution respect for some parts of Maoist ideology, and for figures such as Guan Zhaozhi, seem to show that he was not seriously hurt by the young radicals. He was probably also influenced by his four children, aged 7–11 in 1966. They joined the Red Guards (Hu Zuoxuan and Shi He 2002: 77), and the two eldest daughters 'volunteered' to go to the countryside after high school in 1973–4 (IMCAS 1975).

Years under military rule, 1969–71

The new PLA-dominated establishment did not end the harsh treatment of victims of the preceding period. The institute was now headed by a newly transferred cadre Zhao Weishan 赵蔚山 (1923–87), appointed in November 1969 (Chinese Academy of Sciences 1969), assisted by several PLA-affiliated outsiders. IMCAS was divided into four 'companies' and several 'groups' or 'squads' (IMCAS 1970a). Employees who were not under investigation but still suspect for being old intellectuals had to remain in their offices during the day under surveillance from the Workers Propaganda Team or those Red Guard activists who were co-opted by it. Those who attempted to discuss mathematics could be severely criticized,[13] and most people thus read approved books (essentially Mao Zedong's writings) or chatted about innocuous topics, such as the effectiveness of Chinese medicine in the case of Wu Wen-Tsun and his younger colleague Lu Qikeng, which was sanctioned by Mao's interest in the subject (Lu Qikeng 2010).

Another topic discussed by the confined mathematicians was the theory of relativity. This was initiated by Wu Wen-Tsun, who had read some books about it, and Lu Qikeng considered it legal, because it could be fitted into a movement to criticize the theory as idealist.[14] Eventually, Lu Qikeng was assigned by Guan Zhaozhi to a new Research Section No. 13 controlled by IPCAS, devoted to the criticism of the theory of relativity (Lu Qikeng 2010). Instead of working in a factory, he translated Einstein's German papers into Chinese for the Red Guards, who 'could not read any German and their English was also very bad', so that they could properly denounce Einstein's theory (Lu Qikeng, interview with the author, 7 October 2008).

The leader of the criticism group, a PLA official, rarely visited it, and the researchers thus had a quiet environment to pursue their own interests. Lu Qikeng intended to invite Wu Wen-Tsun to the group too, and Wu seemed eager to study applications of algebraic topology in modern physics. However, this did not materialize, according to Lu Qikeng, for unspecified political reasons. Lu gained familiarity with quantum field theory in IPCAS and used it in July 1972, after a visit by the Chinese-born American physicist C.N. Franklin Yang (Yang Chen-Ning or Yang Zhenning 杨振宁, born 1922), a Nobel laureate. Lu Qikeng attended Yang's talk and was ready to respond to it with an article proving a relationship between Yang's theory and mathematical concepts introduced by Wu Wen-Tsun's teacher S.S. Chern. Wu Wen-Tsun also attended this talk (Lu Qikeng 2008, 2010).

Research Section No. 13 was, however, a rare exception. Most mathematicians could not engage in theoretical topics. Instead, some CPC leaders encouraged renewed focus on high-end technology. For example, a large group of scientists (perhaps close to a hundred) from various CAS institutes, including IPCAS and IMCAS, was dispatched to the Capital Steel Works on

the western outskirts of Beijing, with the task of automating the steel production process. But as none of the researchers had any background in steel
production, they did not have access to literature on the problem, and all
expert metallurgists had been collectively sent down to the countryside, the
movement produced no results in the six months of its duration (Du Junfu
2009a: 30). IMCAS researchers also engaged in other automation tasks in
1970 (IMCAS Revolutionary Committee 1971).

In order to remould themselves more quickly, all researchers periodic-
ally went into the 'production practice'. This was one of the 'socialist new-
born things', hailed as a major breakthrough of the Cultural Revolution.
Virtually all institutions with intellectuals and other cadres established *May 7
Cadre Schools* (*Wu Qi ganxiao* 五七干校), located in the countryside, where
employees of all ranks engaged partly in agricultural work and partly in political study. Conditions in these 'schools' varied from rough to harsh.[15] The
CAS May 7 Cadre school was located in Hubei province. Of the 272 IMCAS
employees, 90 passed through the institution in four groups in 1970, including Chen Pihe, Wu Wen-Tsun's wife, who worked in IMCAS as a librarian
(IMCAS 1970a, 1970b). The third and fourth groups spent more than a year
in Hubei: the last IMCAS employees were still in the school in December
1971 (CAS May 7 School 1971). In December 1970, 80 people were at the
school and 73 in various mines and factories.

Wu Wen-Tsun did not go to a May 7 school, but instead directly to the First
Wireless Telegraphy Factory in Beijing (*Beijing Wuxiandian yi chang* 北京无
线电一厂), where he also spent several months from the summer of 1970
until 1971 (Wu Wen-Tsun 1976; IMCAS 1975). In later articles and interviews, Wu emphasized how this encounter with computers made him aware
of their power (Hu Zuoxuan and Shi He 2002: 76; Yang Xujie 2005), but in
1976 Wu only claimed that this experience allowed him, apart from learning the 'proletarian spirit' from the workers, to make his earlier applied work
truly applicable. His method for producing planar realization of graphs was
allegedly helped by the work in the factory (weld testing). He became aware
of the practical implications of the theory, and adapted his method to suit
them. He realized that 'similar foreign works are not only theoretically defi-
cient, but their methods are also incomplete, and our work is in comparison
much better' (Wu Wen-Tsun 1976: 14).

Wu Wen-Tsun also studied Mao Zedong's works *On Practice, On
Contradiction* and *Speech at the Yan'an Forum on Literature and Art* (Wu
Wen-Tsun 1976).[16] Direct study of Mao Zedong's works could be a relatively
refreshing experience, seen in contrast to the pre-1966 'political study' sessions, weekly lectures by the lowest-level cadres on the recently issued central
directives and *People's Daily* editorials.

In 1976, Wu Wen-Tsun spelled out how the study of Mao Zedong's Thought
had helped his research:

> Many geometric and mathematical disciplines require at some point of
> their development a unifying approach or style in order to be generalized.

In the West, several such styles have been suggested, for instance the unification of geometry based on transformation groups, or unification proceeding from axiomatic systems. (...) We do not want to discuss the relative merit and philosophical problems of these foreign doctrines. On the basis of Marxism–Leninism, Mao Zedong Thought and with reference to our ancient mathematics, we propose a different unifying style for topology, centred around the concept of measure.

> (Wu Wen-Tsun 1976: 15–16)

He formulated it more pointedly in 2003:

I studied many of Chairman Mao's works during the Cultural Revolution. I feel that several of Chairman Mao's points about war provide a certain guidance for science. Chairman Mao talked a lot about how to turn a disadvantage into an advantage. At the time I wondered how to turn the disadvantage of Chinese mathematics into an advantage? Traditional Chinese mathematics gave me inspiration right at this time.

> (Jiang Yan 2003: unpaginated)

Wu's understanding of the concrete contents of Mao's dictum is revealed from this passage in a TV interview (the militant facial gesture at the end sadly cannot be conveyed in writing):

Several years after the establishment of the Institute of Systems Science, when mechanical proofs where accepted even abroad [this happened in the mid-1980s], things like the following could still happen in China. At a conference, I encountered someone from IMCAS – an important person, I don't want to say anything more. And he criticized me: 'Everyone abroad does mechanical proofs by mathematical logic. Why do you not use mathematical logic?' At this point I immediately got angry, I started to openly quarrel with him. (...) I said – if the foreigners do mathematical logic, do I have no option but to do mathematical logic? 'What the foreigners do is what I won't do, what the foreigners don't do is what I will do'. That's my basic principle.

> (Li Xiangdong and Zhang Tao 2006: 40:15–41:15)

This strategic thinking was produced not only by immersion in Mao's theoretical works, but undoubtedly also by the necessities of self-preservation in a turbulent political climate. The constant struggle meetings, armed fighting between Red Guards, the PLA's involvement in all civilian affairs and increasing war paranoia generated by the escalation of the Vietnam war and Sino-Soviet tensions all generated a sense of personal insecurity and a need to be circumspect and think strategically. On a deeper level, all official discourse in this period was couched in war-like terms, and the entire Cultural Revolution was presented as a fight against a series of cunning, vicious conspiracies

(Walder 1991). Wu Wen-Tsun, an introvert and avid chess player (Li Banghe 1989), responded by becoming fond of strategic thinking.

Fear and mental stress peaked in the years 1969–71. The mathematical physicist Zhang Zongsui 张宗燧 (1915–69), branded as 'bourgeois academic authority', committed suicide in June 1969, unable to stand the continuous repression and harassment. The 76-year-old Hiong King-Lai 熊庆来, Hua Loo-Keng's teacher and benefactor, who had returned from France only in 1958, died of heart failure at his home in early 1969, after a prolonged period of denunciations and forced self-criticism for his role in the pre-1949 KMT establishment (Wang Yuan 1999: 338–51).

The new PLA-dominated leadership at CAS, however, also targeted the former radicals in the nationwide 'May 16 Conspiracy Investigation'.[17] Two IMCAS radicals, Zou Xiecheng and Xu Mingwei, figured prominently among the suspects. The main issue was the illegal seizure of power in CAS against the wishes of Zhou Enlai in 1967, theft of top-secret archives, and perhaps most importantly their collusion with higher-level radicals in questioning the authority of Zhou. Publications of the CAS radicals had accused the Premier in 1967 – in allegorical yet unambiguous terms – of blocking the Cultural Revolution (IMCAS 1973c). Zou Xiecheng was kept in confinement, and died during the investigations, allegedly by his own hand, on 17 July 1970 (Chinese Academy of Sciences [CAS CPC organization] 1981).[18] Xu Mingwei survived the Cultural Revolution (on a salary reduced to 45 yuan) and his penalties were lifted in 1981 on the basis that his mistakes were only due to having been cheated by the Gang of Four (IMCAS Party Steering Group 1981). More fortunate suspects were criticized in small-group meetings and forced to make confessions. This also affected Guan Zhaozhi for his involvement with the radical Red Guard faction in the previous period. Although he retained some influence and ideological authority in the 1970s, he was not a member of the new IMCAS leadership, and there was a discretely conducted investigation of his 'special case' (Wang Yuan 1999: 351).[19]

A large part of the 'May 16 Conspiracy' investigation was based on confessional or extorted denunciations, which produced lasting distrust and strained personal relationships in IMCAS. The mistakes in the conduct of the investigation were acknowledged in the final report (IMCAS Steering Group 1975b), which also stipulated that all 'evidence' related to remaining suspects should be destroyed or, in the case of confiscated personal documents, returned to their owners. For this reason, there is also no record of the number of people affected.

The situation in IMCAS during the years 1966–71 should be put into perspective by a comparison with the fate of CAS as a whole. During the first two years of the CR, all seven top CAS officials, 59 of 71 section-level officials, 99 of 192 minor officials, and 131 of 170 Beijing-based senior researchers were dismissed and persecuted. Altogether, 106 Beijing employees of the CAS (0.41 per cent) died because of political persecution between 1966 and 1976. The situation in IMCAS was more serious: at least five deaths related to political persecution occurred in 1966–76, four suicides and one natural death during house arrest.[20] This represented 1.84 per cent of all IMCAS employees, significantly above the CAS average.

The number of CAS research institutes dropped from 106 in 1965 to 13 in 1973. Forty-three were governed jointly with other agencies, mostly PLA-related, and 47 were handed over to the PLA-administered Commission for New Technologies. Regional authorities took control of many institutes, factories and libraries, and often regarded them as a useless burden. CAS Beijing Botanic Garden was closed down and converted to vegetable plantations. A large part of research institutes' material equipment was destroyed or damaged, either intentionally or by neglect (Chen Jianxin, *et al.* 1994: 228–39).

The Cultural Revolution also led to an almost complete suspension of international contacts in 1966–71. Chinese students were expelled from the Soviet Union and other East European countries in late 1966 for their radical provocations, and students from all other countries (including France) returned when the Ministry of Higher Education, which had been seized by rebel Red Guards, called on them to come home and join the Cultural Revolution (Chen Jianxin, *et al.* 1994: 255–6). China cultivated diplomatic relations only with Albania during this period.[21]

The late CR period, 1971–78

The situation produced by the violent phase of the CR period was not stable. It was clear that Mao Zedong would stay in charge until his death, and the issue of who would replace him generated conflicts between the different constituencies that were brought together to govern post-CR China. Mao's chosen successor, Marshal Lin Biao, was quickly losing the Chairman's favour in 1971, and then shockingly died with his wife and sons in a plane crash on 13 September, probably after a failed attempt to flee to the Soviet Union. This ended the PLA's domination of Chinese politics, and persuaded Mao that the country should return to a more civilian form of government. The extent to which the changes brought about by the CR should be preserved was, however, a point of controversy between the pre-1966 cadres and new leaders recruited among the CR radicals. The staging of this conflict was ultimately controlled by the increasingly frail and inaccessible Mao, whose support switched between various figures in both camps. In the wider society, the conflict was reflected in alternating campaigns of consolidation and renewed ideological struggle, influencing policies and personnel decisions at all levels.

The conflict finally had to be resolved in 1976, when both Zhou Enlai and Mao Zedong died. Although the radicals found themselves isolated after Mao's death and were purged and arrested in October 1976, it was far from clear which way the country was going to move. Mao's successor Hua Guofeng was a pragmatist, but his legitimacy was derived from Mao's and he could not therefore radically break with the CR past. However, criticism of the purged radicals undermined his position, and he had to gradually cede power to Deng Xiaoping, whose programme of modernization and reform, formulated in 1978, finally marked the end of the CR period. Reassessment of science and technology was an important component of Deng's new line.

This section will provide an outline of the Institute of Mathematics' development during this period. Two issues of particular relevance to Wu Wen-Tsun – international cooperation and the turn to historical studies – will be analyzed in subsequent sections in more detail.

As can be seen from Table 4.2, a change of direction occurred almost every year. This creates a difficulty in assessing contemporary sources: the IMCAS leadership compiled annual reports in this period, but they were often written just as last year's policies had been rescinded and replaced by their near opposites. Serious distortions and change of emphasis in them can consequently be expected.

The death of Lin Biao on 13 September 1971 was a symbolic turning point in the history of the CR period.[22] It hastened the restoration of civilian order (to curb the PLA's power) and rapprochement with the capitalist countries. These processes were already under way in 1970–1, but the brutal fact of Lin Biao's defection and death forced Mao Zedong and the CPC leadership to change direction more radically. It took some time before the incident was revealed to the public, but by the end of 1971 Zhou Enlai, the Premier, had gained Mao's permission to correct some of the excesses attributed to Lin Biao.

Zhou allowed many disgraced cadres to return to office (MacFarquhar and Schoenhals 2006: 339–54). Re-investigations of the cases of some of the CR victims (roughly 30%) in IMCAS included the former Deputy Director and Secretary of the Party Committee Zheng Zhifu, as well as another old CPC member Sun Keding (IMCAS 1972b). Cadres still in the 'May 7 Schools' were sent back to their work units, and senior researchers were no longer forced to spend a long time working manually outside their institutions.[23]

Older, experienced cadres with higher regard for intellectuals joined the radicals in the revolutionary committees. The inner organization of IMCAS returned to the pre-1966 model, with discipline-related research sections instead of the 'squads' of 1970. The theoretical subjects (number theory, algebra, topology, functional analysis and function theory) were combined into a 'Five Subjects Research Section', alongside sections for the study of partial differential equations, probability and statistics, operations research, and control theory, a computation centre, and a computer factory (Fitzgerald and MacLane 1977: 11). Zhao Weishan, who became the head of the CPC branch in 1969, was seconded by Tian Fangzeng as Deputy Director. (Hua Loo-Keng formally remained director but in fact avoided IMCAS.) Tian, a French-educated mathematician and CPC member, also became the head of the Five Subjects Section.

Other signs of relaxation included resumed publication of scientific journals. *Scientia Sinica*, publishing articles in foreign languages, was the most

Table 4.2 Principal stages of the political conflict between 1971 and 1978

Stage and Time	Description
Post-Lin Biao Consolidation 1971–late 1973	Influence of the PLA diminished, restoration of civilian order and relaxation of some ideological requirements, renewed contacts with the West. Premier Zhou Enlai in charge.
Anti-Lin Biao Anti-Confucius Campaign 1973–summer 1974	Radicals in the Politburo persuade Mao that consolidation had gone too far to the right. Historical allegories used to criticize attempts to reverse some CR policies and verdicts, aimed against Premier Zhou Enlai.
Deng Xiaoping's Consolidation late 1974–October 1975	Disruption caused by the previous campaign, poor performance of radical leaders, and Mao's and Zhou Enlai's declining health force Mao to appoint a pragmatist, Deng Xiaoping, as acting Vice-Premier. Improvement of the economy, restoration of disgraced cadres, increasingly bold measures also proposed for science and technology, emphasis on Four Modernizations.
Repudiating the right-opportunist wind to reverse the verdicts February–October 1976	Mao loses confidence in Deng and initiates a campaign against him, first in the top leadership, from February 1976 also at mass rallies. Popular discontent with the radicals apparent after Zhou Enlai's death (Tiananmen Incident, April 5). The devastating Tangshan earthquake (July 21) and Mao's death (September 9) frustrate the radicals' effort to focus on the anti-Deng campaign.
Smashing the Gang of Four October 1976–July 1977	Zhou Enlai's and Mao Zedong's successor Hua Guofeng decides to remove the Gang of Four in a military coup. Hua starts a campaign (again with mass rallies) to criticize the Gang, initially resists the return of Deng and reaffirms the CR, but the criticism of the Gang grows deeper and leads to Deng's rehabilitation in the summer of 1977. Scientific research starts being supported.
Rehabilitation of victims, 'Spring of Science' July 1977–April 1978	A complete reversal of CPC's long-standing approach to intellectuals, who are now accepted as part of the working class. Science and technology are declared to be 'productive forces', international exchange intensifies.

Source: According to Teiwes and Sun (2007).

important, but *Acta Mathematica Sinica* also started to publish long-delayed research results. New journals were created too. *Mathematics in Practice and Theory* (*Shuxue de shijian yu renshi*) was started by IMCAS for restricted circulation in 1972 to emphasize the new connection between theory and practice achieved by the CR. Debates about the correct research direction of IMCAS were organized in late 1971 and early 1972. 'The academic atmosphere gradually became more lively, we gained preliminary understanding of developments in world mathematics, and seminars or conferences were held almost every day' (IMCAS 1973a). Universities started enrolling students again, although only among workers, peasants and soldiers, and on recommendation rather than through entrance examinations, which caused resentment and invited corrupt behaviour (MacFarquhar and Schoenhals 2006: 371).

In May–July 1972, the State Council under Zhou Enlai organized a large meeting on educational policy, which emphasized the need for more attention to fundamental theory (as opposed to practical applications) in university education. Zhou Enlai asked the famous physicist and President of Peking University, Zhou Peiyuan 周培源 (1902–93), to write an article on the topic, which was published in October in the national daily *Guangming ribao* (Teiwes and Sun 2007: 59–60). Premier Zhou's directives became an object of close reading at IMCAS (IMCAS 1973a).

It was in this period of relaxation that Chen Jingrun, one of the most widely known modern Chinese mathematicians,[24] suddenly became famous. Chen had come to IMCAS in 1956 and joined Hua Loo-Keng's number theory seminar. His extremely introverted character and single-minded attention to mathematics made life difficult for him in all political movements. Shortly before the Cultural Revolution, Chen announced a proof of a theorem hitherto closest to the Goldbach conjecture, called in short '1+2'.[25] The proof was not published, however, because a reviewer (Min Sihe 闵嗣鹤 from Peking University) recommended that it be simplified. During the Cultural Revolution, Chen was ridiculed as a 'white expert' and suffered a lot of harassment, but continued to work slowly and laboriously on a better proof of his theorem, despite appalling living conditions, poor health and constant fear of criticism.

In autumn 1972, the Five Subjects Research Section, where Chen was based, received a new CPC branch secretary. Li Shangjie 李尚杰 (born *c*.1930) was a political commissar from the PLA, but he held intellectuals in high esteem and became especially protective of Chen, who gradually developed a strong trust in him. In the spring of 1973, Li persuaded Chen to release his proof for publication. Li promised Chen that his paper could not bring him any harm, however theoretical it was. Then a PLA general visiting IMCAS pressured the leadership of IMCAS to let Chen publish his article, which eventually appeared in *Scientia Sinica* (Chen Jingrun 1973). Shortly afterwards, Chen's result was praised by the Vice-President of CAS, Wu Heng, and in late April 1973, Chen's success and his less than ideal living conditions were raised (against opposition from the IMCAS Revolutionary Committee) by two reporters of the New China News Agency in their restricted-circulation bulletin. These articles slightly overstated Chen's health problems, prompting Mao's wife Jiang Qing and Mao himself to order Chen Jingrun's immediate transfer to a hospital for the best available treatment. Chen Jingrun even became a delegate to the 1974 National People's Congress (Xu Chi 1978; Shen Shihao 1997; Wang Lili and Li Xiaoning 1998; Luo Shengxiong 2001).[26]

Within IMCAS, Chen Jingrun was already praised by name in 1972, albeit in secondary place to others' applied, practically oriented, results. The work report highlighted the application of Wu Wen-Tsun's work on the layout of integrated circuits to a debugging program for an automatic IC layout machine, and Hua Loo-Keng's popularization of operations research. Even replies to letters from the public and popularization brochures were used as evidence of the new socialist research orientation (IMCAS 1973a).[27] Less conspicuous but equally important was military research, for example Wan Zhexian's work on coding and decoding problems since 1971 (Interview with the author, Beijing, 7 October 2008). Theoretical subjects were tentatively explored in 1973, including several topics in topology and differential geometry by Wu Wen-Tsun and his students Li Banghe, Li Peixin, Yu Yanlin and Wang Qiming. Rhetorical emphasis on applications was, however, still present, as reflected in Wu Wen-Tsun's project 'Applications of geometry and topology in celestial mechanics, general relativity, high-energy physics, biology, etc.' (IMCAS 1973b)

Ideological relaxation following the Lin Biao Incident did not last very long. In December 1972, Mao reappraised Lin Biao as an ultra-rightist, and the IMCAS 1972 work report, written in 1973, consequently criticizes Lin Biao's rightist errors. He was reported to have denigrated the May 7 schools as 'masked labour camps' and sending down of youth as 'masked unemployment' (IMCAS 1973a). From mid-1973, radicals started actively attacking the policies promoted by Zhou Enlai, including Zhou Peiyuan's article about increasing fundamental research. The new university entrance examinations, introduced to ensure a minimal level of education for new students, were attacked by an unsuccessful candidate in 1973 as directed against the workers and peasants, who had served the people instead of reading books. Zhou Enlai was further frustrated by Mao's clear preference for the young radical leader, Wang Hongwen, and towards the end of 1973, a series of Politburo criticism meetings was organized against Zhou by Mao. In January 1974, the radicals launched a major 'Campaign to criticize Lin Biao and Confucius' (*Pi Lin pi Kong yundong* 批林批孔运动), which they hoped would be 'a second Cultural Revolution' (Wang Hongwen) and cement both the results of the first revolution and their position at the expense of Zhou Enlai (Teiwes and Sun 2007).

Unlike the actual Cultural Revolution, the Anti-Lin Anti-Confucius campaign was conducted chiefly through propaganda and mass meetings organized by the new power holders rather than spontaneously. Even this created considerable disruption and uncertainty. The IMCAS work summary for

1973 is not registered in the archives at all, and the report for 1974 was only produced after the Anti-Lin Anti-Confucius campaign lost vigour and Mao's emphasis switched to consolidation. The report thus said very little about the campaign and focused on uncontroversial positive achievements in learning Marxist–Leninist theory and applying theoretical knowledge into practice. The Probability and Statistics Research Section was praised for being a model in the Anti-Lin Anti-Confucius campaign as well as in other politically laudable activities, such as popularization (IMCAS Steering Group 1975a).

Starting in July 1974, Mao was already limiting the scope of the campaign once it had achieved his goal of humiliating Zhou Enlai. Mao's favour shifted towards Deng Xiaoping. Although Deng had been heavily criticized in 1966–8, he was allowed to return to the government in 1973 and eventually became First Vice-Premier and Chief of Staff. From January 1975, he controlled most areas of policy and, with Mao's endorsement, suppressed radical disruption and factionalism.

In the first half of 1975, Deng focused on non-problematic areas such as railways. The 'superstructure', including science and technology, was still mostly in the hands of the radicals, who organized further propaganda campaigns to study the theory of proletarian dictatorship and to combat empiricism (ostensibly but not explicitly against Zhou Enlai). Mao grew more displeased with the radicals and reprimanded them for behaving like a Gang, which allowed Deng to expand his consolidation efforts to include the Chinese Academy of Sciences. Working under Deng, Hua Guofeng, Hu Yaobang (later CPC Secretary General) and Li Chang came to CAS to understand its problems and suggest improvements, eventually drafting *The Outline Report on the Academy of Sciences* (*Kexueyuan gongzuo huibao tigang* 科学院工作汇报提纲). They effectively sidelined the radical leadership and strengthened research at the expense of politics (Teiwes and Sun 2007; Chen Jianxin, *et al.* 1994: 248–51). IMCAS compiled a 'situation report' in July 1975, probably for this 'work team' of sorts. The report acknowledged mistakes in neglecting the talent of Chen Jingrun, but still suggested that abstract topics like his could not be pursued by too many people. The report also complained about shortage of office space, lack of young researchers (the youngest were 34 years old), and the fact that one third of researchers could not live together with their spouses because of different employers (IMCAS Steering Group 1975b).

The year 1975 was perhaps the most relaxed and productive of the entire period before Mao's death. Wu Wen-Tsun did particularly well, writing and publishing a host of papers on his new theory of the I^*-functor, and visiting the Institut des Hautes Études Scientifiques in Bures-sur-Yvette in France in May and June, with his student Wang Qiming and another older topologist Zhang Sucheng (IMCAS 1975; IHÉS 1975). The ideological campaigns, such as the study of the theory of proletarian dictatorship and the study of the novel *The Water Margin* (*Shuihuzhuan* 水浒传), still continued, but limited time was devoted to them and they did not stray into direct criticism of living people.

But scientists remained cautious about the prospects of Deng's consolidation, with good reason. In late 1975, Mao Zedong became convinced that Deng was dissatisfied with the Cultural Revolution, and approved a campaign against him and his policies, which gradually escalated and led to Deng's second dismissal from all offices in late January 1976 (MacFarquhar and Schoenhals 2006: 404–16). In February, criticism of Deng's mistakes moved past the inner CPC circles and a new campaign 'to counter-attack the right-opportunist wind to reverse verdicts' (*huibi youqing fan'an feng yundong* 回击右倾翻案风运动) was launched with mass meetings.

The complexity of the changing situation can be seen from the IMCAS 1975 work report. One of the measures suggested by Deng's men in CAS was to limit 'open-door research' (*kai men ban keyan* 开门办科研), i.e. working in the wider society in order to 'link theory with practice'. But radicals in the applied disciplines put up strong resistance:

> Many comrades have already become addicted to the sweet taste of open-door research and are determined to walk further down this road. For a certain period, the advocates of the right-opportunist wind to reverse verdicts in science and technology circles attacked open-door research, unfoundedly claiming that 'when there is too much linking with practice, theory is suppressed', 'the fewer such innovations the better' and so on. They foolishly tried to return our science and technology workers back to the pre-CR revisionist road, to shut them in secluded buildings and aloof academies, and cut them off from the bustling affairs of socialist construction. This revisionist line that they advocated deservedly met with resistance from the broad masses of revolutionary intellectuals.
>
> (IMCAS Steering Group 1976)

The leadership humbly self-criticized for allowing this 'revisionist nonsense' to go unchecked. Its excuse was that 'Hu Yaobang and Li Chang found in our academy a market for the revisionist absurdities they advocated, which limited our ability to resist them' (IMCAS Steering Group 1976).

We have an unusually complete record of the campaigns of 1976 and 1977 in 37 brief situation reports, written by the Institute leadership after every major rally, sometimes even twice a day. They show that the campaign to criticize the 'right-opportunist wind' was orchestrated by the leadership, with Zhao Weishan, the Deputy Director and Head of the Steering Group, appearing particularly vitriolic. Most speakers at rallies hailed from the worker–peasant–soldier members of the applied research sections and the affiliated computer factory.[28] By March, Deng Xiaoping was being criticized by name. A major turning point was the Tiananmen Incident in April, when Beijing residents turned commemoration of Zhou Enlai, who had died in January, into a demonstration of their discontent with the radicals in the Politburo, particularly

Mao's wife Jiang Qing. Tiananmen Square was forcibly cleared and a search for perpetrators began. IMCAS also reported some offences, such as sending wreaths after an official ban, and 'spreading rumours' (IMCAS CPC Steering Group 1976b). Some people were reported to the CAS security department, but most were criticized within IMCAS. Criticism continued throughout May, but refocused on praising the CR's results in science and technology and attacking Deng's revisionism. Wu Wen-Tsun made a speech on May 15 at a rally commemorating the tenth anniversary of the May 16 Circular, which had started the Cultural Revolution (IMCAS CPC Steering Group 1976c).

The campaign slowed down in the summer, especially in August after the Tangshan earthquake of 21 July 1976. Subsequent meetings emphasized disaster relief rather than political struggle. In early September, Zhao Weishan reported that further escalation of the anti-Deng campaign had met with considerable resistance, as some researchers complained that it had gone on for too long and was becoming futile, some that linking theory with practice was just too difficult, and some that they were being robbed of valuable research time. Zhao decided to counter these problems by focusing on the Five Subjects Research Section and criticizing its failure to engage in open-door research (IMCAS CPC Steering Group 1976d).

Three days later, Mao Zedong died. The rest of September was devoted to eulogizing, with a major meeting on 22 September. Guan Zhaozhi, Wu Wen-Tsun and Chen Jingrun all delivered eulogies. Guan remembered meeting Mao in 1950, while Wu emphasized how Mao cared about the correct thought of intellectuals and their political education:

> [Wu Wen-Tsun] is determined to turn the grief into a strength, continue to study Mao Zedong's works, take class struggle as the key link and actively participate in the criticism of Deng and revisionism, in order to train himself in real-life struggles. He is determined to continue to arm his brain with Mao Zedong Thought, try to make it guide his research work, and implement it in real life. He is determined to rely on Chairman Mao's enlightened directives about science and technology, and to develop mathematics according to Mao Zedong Thought, using all his powers to elevate our mathematics to an unprecedented summit in cooperation with other comrades.
>
> (IMCAS CPC Steering Group 1976e)

The stream of brief reports then stops until November. Presumably the interpretation of current events changed too quickly and the reports were destroyed (some reports are also missing for earlier crucial moments, judging by gaps in numbering). On 10 October, the Gang of Four was arrested, and Hua Guofeng started a new campaign against the radicals. This extended to their chief assistant in the CAS, Liu Zhongyang 柳忠阳. Criticism meetings were taking place every week, and previously reluctant or sceptical researchers suddenly participated with much more enthusiasm.[29] From the start, Zhao Weishan was under heavy pressure for failing to lead the movement as actively as he had led the anti-Deng campaign, and by the end of December he was writing self-criticisms and fighting for his managerial life (IMCAS CPC Steering Group 1976g). After some respite around the 1977 Chinese New Year Holiday,[30] criticism of Zhao Weishan intensified, and he was suspended from his posts and made to write self-criticisms in isolation.

The rest of 1977 and 1978 were marked by the final ascendancy of Deng Xiaoping and marginalization of Hua Guofeng. For scientists, the period was a true liberation after the CR movements, and the whole series of political campaigns since at least 1957. The CPC called a Science and Technology Conference to meet in Beijing in April 1978 (Yuan Zhendong 2008), and positive reports about Chinese science and technology experts flooded national newspapers in the lead-up to the conference. Mathematicians were among the first to be celebrated. The short story about Chen Jingrun, published in the magazine *People's Literature* in January 1978 and reprinted in the *People's Daily* shortly afterwards, had an especially profound impact on the perception of scientists in post-Mao China (Sun Wenhua 2008).

The political struggles of the late CR period were exhausting and frustrating, but much more orderly than the Cultural Revolution itself. In a way, they were also much more engaging for larger numbers of people, because the clear dividing line between the Red Guards and the old intellectuals disappeared. The result was a higher level of independent thought on matters broader than one's own specialization; people also became more articulate in expressing them. This became clear in 1976 and 1977, when many researchers were able to resist or deflect the anti-Deng campaign, and later actively expand the criticism of constraints imposed by radical ideologues.

These positive comments should not overturn the basic verdict on the CR period in Chinese mathematics: it was a major disruption, which cost many researchers the best years of their career and wasted a lot of opportunities, not to mention the personal tragedies it caused. But it also created unique inspirations and injected those who were not broken by the times with a hard-working spirit and optimism after 1978.

World mathematicians in China in the 1970s

Apart from these domestic developments, the late CR period also witnessed the end of Communist China's diplomatic isolation. The People's Republic established and gradually expanded contacts with Japan and the USA (President Nixon famously visiting Beijing, Hangzhou and Shanghai in February 1972), and gained a UN seat at the expense of Taiwan in 1971. Visitors from the USA, Japan and Europe were coming in numbers unseen since 1949. An important component of these international exchanges, prominently reported in national newspapers, was science. New international contacts benefited some senior members of IMCAS in particular, who were able

Table 4.3 Visits of mathematicians reported in the *People's Daily*, 1972–76

Members	Country, discipline	Date of visit
Thomas E. CHEATHAM Jr., Wesley A. CLARK, Anatol HOLT, Severo M. ORNSTEIN, Alan J. PERLIS, Herbert A. SIMON	USA, computer science	July 1972
Hao WANG 王浩, S.S. CHERN, Chia-chiao LIN 林家翘, Hsien-chung WANG 王宪钟	USA, mathematics	July 1972
Kai-lai CHUNG 钟开莱	USA, mathematics	Summer 1972
Donald C. SPENCER, William BROWDER, Franklin P. PETERSON	USA, topology	April–May 1973
Nicolaas H. KUIPER	France, mathematics	April–May 1973
Jacob T. SCHWARZ	USA, computer science	July 1973
Chia-Chiao LIN	USA, applied mathematics	October 1973
Dan MOSTOW	USA, mathematics	1974
Hao WANG	USA, mathematics, computer science	December 1973 –January 1974
S.S. CHERN	USA, mathematics	Sep–Oct 1974
Louis NURENBURG	USA, mathematics	November 1975
Chia-Chiao LIN	USA, applied mathematics	April–May 1976
S. MACLANE, J. KOHN, E. BROWN, G. CARRIER, W. FEIT, J. KELLER, V. KLEE, H. POLLAK, H.H. WU 伍鸿熙	USA, mathematics	3–27 May 1976
André WEIL	USA, mathematics	6–22 October 1976

Sources: *People's Daily* 1972–76.

to meet foreign mathematicians, and occasionally even leave the country (Hua Loo-Keng visited Japan in 1972).

A recent book (Liu Qiuhua 2010) summarizes the most important mathematical exchanges described in other published sources. The first Western mathematician to visit the PRC was apparently Chandler Davis (born 1926) from the University of Toronto. The physicist C.N. Yang also visited China in 1972. The importance of this visit and C.N. Yang's lectures for Lu Qikeng have already been mentioned above, but the visit also had a political significance. It was Yang who, in conversation with Zhou Enlai, brought up the question of elevating the level of fundamental research, and this inspired Zhou Peiyuan's article, later criticized by the radicals (Lu Qikeng 2008).[31] Moreover, in 1973, a group of Chinese students, including the mathematician Yan Jia'an (严加安, born 1941), was sent to France for more than two years of intensive training. Yan studied probability at the University of Strasbourg (Cheng Minde 2002: 637–48).

Mathematicians and computer scientists whose visits were reported in *People's Daily* are shown in Table 4.3.

The visits of Wu Wen-Tsun's former teacher S.S. Chern were especially long, frequent and important for raising Chinese mathematicians' awareness of mathematical developments. IMCAS organized a collective study of topological topics necessary for a proper understanding of Chern's work (Peng Jiagui and Hu Sen 2010). Other Chinese-born mathematicians also played a role in mediating between Chinese and world mathematics. The most significant for Wu Wen-Tsun was Hao Wang, a mathematical logician, who had introduced the concept of mechanization of mathematics (Wang 1960), a point noted by Wu (Wu Wen-Tsun 1980). Wang later wrote a preface to the volume where Wu's articles were first introduced to the Western mathematical community (Bledsoe and Loveland 1984).

Among non-Chinese mathematicians, the delegation of US topologists in April 1973 brought to China papers by the American topologist D. Sullivan (born 1941), which inspired Wu Wen-Tsun's work on the I^*-functor (Hu Zuoxuan and Shi He 2002: 77–8). The only French visit by the (Dutch) director of IHÉS Nicolaas Kuiper in 1973 probably enabled Wu and his colleagues to visit IHÉS in 1975.

The largest and most famous visit was the American Pure and Applied Mathematics Delegation in May 1976. It published an extensive report of its findings, describing the current situation in Chinese mathematics, especially the extreme attention to applications. It occurred at the height of the campaign against Deng Xiaoping, in May 1976. Wherever they went, the members of the delegation would always first be told about the campaign and its significance (Fitzgerald and MacLane 1977: 6). Although the delegation reported favourably on some aspects of Chinese mathematics, especially the work of Chen Jingrun, Wu Wen-Tsun, and Hiong King-Lai's students Zhang Guanghou 张广厚 and Yang Le 杨乐 (two relatively young mathematicians working on the theory of meromorphic functions), its overall assessment was rather worrying. Chinese mathematics lacked a substantial inflow of young researchers, was managed by administrators without experience with 'first-rate creative work', and was oriented towards 'repeated application of known techniques' instead of new challenges from science and technology (Fitzgerald and MacLane 1977: 68–9).

The campaigns of the 1970s did not disrupt international visits, possibly because such visits often took one to two years to organize (CSCPRC 1974). International guests had special privileges, and could sometimes even help troubled Chinese colleagues. For example, Lu Qikeng was not allowed to see foreign guests in 1976, but H.H. Wu specifically requested to meet him, and provided him with a unique opportunity to discuss his work (Lu Qikeng, interview with the author, 8 October 2008).

Wu Wen-Tsun was inspired by these renewed contacts to return to algebraic topology, although his views on mathematics had already substantially shifted. This was reflected in the unusual character of his new research agenda, inspired by the theory of minimal models of differential graded algebras due to Sullivan (1973). Wu familiarized himself with the relatively new language of category theory, and reformulated Sullivan's minimal models as a functor (called the *I**-functor) from the category of locally finite simply connected simplicial complexes *W* (in later articles expanded to the category of all connected topological spaces) to the category of differential graded algebras *A*. The use of the functor concept emphasized a correspondence between geometrical and algebraic aspects of mathematics, linking conveniently to Friedrich Engels' view of mathematics as a 'science of spatial forms and quantitative relations in the real world'. Wu often referred to Engels, such as in the following introduction to his first short note (Wu Wen-Tsun 1975b), repeated in a longer English article (Wu Wen-Tsun 1975c):

> Mathematics has for its objects of study space forms and quantitative relations in the [external] world. For algebraic topology in particular, the space forms involved are the various kinds of topological spaces. Its main tools and methods consist of making correspondences usually called functors [between] such spaces [and] certain kinds of quantities expressed in terms of algebraic structures [such] as numbers, groups, rings, algebras, etc., by means of which properties of spaces and also their variations can be investigated.
>
> (Wu Wen-Tsun 1975c: 464, corrected according to the Chinese version in Wu Wen-Tsun 1975a and Wu Wen-Tsun 1975b)

Wu noted that Sullivan's minimal models are calculable for a much wider range of situations than more traditional tools of algebraic topology such as homotopy or cohomology groups. The 'calculability' (*ke jisuan xing* 可计算性) Wu sought meant the possibility to express the minimal model of a space as a formula of known minimal models of spaces from which the given space could be constructed (union, product, etc.). This meant a stronger connection from 'space forms' to 'quantitative relations' and, by implication, to the real world:

> From ancient time, people have expressed certain properties of shapes by numbers, i.e. the measures of shapes, such as the length of roads, the areas of fields, the capacity of vessels, all the way to the cardinality of sets, etc. (...) These two fundamental notions of mathematics [forms and quantities] are, however, not to be considered unrelated, but are often interconnected by 'measures'. Previously we have introduced the concept of *I** which serves as a measure of space forms by means of quantitative relations. This measure is called a 'functor' to follow the current terminology in algebraic topology. (...) Because of the 'calculability' of the *I**-functor, it is easier to master than traditional functors, and we expect it to gain wide application.
>
> (Wu Wen-Tsun 1975a: 162)

The publication of this last article, finished in April 1975, had to wait until October 1976. The Chinese version was delayed even longer (Wu Wen-Tsun 1977a). Despite Wu Wen-Tsun's philosophical effort, work on topology was again not welcome as radicals attacked Deng Xiaoping and his policies in research and education. This atmosphere persisted until the end of 1976, and set the tone of Wu Wen-Tsun's article about his transformation during the Cultural Revolution (Wu Wen-Tsun 1976).

Wu's work on the I^*-functor continued after the Cultural Revolution. He wrote another paper with his student (Wu Wen-Tsun and Wang Qiming 1978), presented the topic at several conferences, and gave a lecture course on the subject at Berkeley in 1981. Expanded lecture notes from this course were published by Springer as the book *Rational Homotopy Type* (Wu Wen-Tsun 1987c).

Although all of this work appeared in English, only Wu Wen-Tsun (1975c), which proved a formula concerning the minimal model of a fibre square, generated citations (3 in MathSciNet, 6 in Google Scholar and 7 in the Web of Science). Most of the citations were in the papers of Chen Kuo-Tsai (Chen Guocai 陈国才, 1923–87), S.S. Chern's student, who moved to Taiwan and eventually California after 1949. Chen reviewed Wu's first paper of the series (Wu Wen-Tsun 1975a) for *Mathematical Reviews*.

Western readers of Wu's work noted that although he was not the first to prove his theorems, his proofs were 'both direct and elementary' compared to earlier alternatives (Lemaire 1979). However, they also complained about Wu's non-standard or obsolete concepts, and repetition of well-known results (Lemaire 1979, 1981; Stasheff 1982; Arkowitz 1989).

Wu Wen-Tsun himself felt strongly the huge challenge in catching up with his rapidly advancing subject. He later recalled:

> The materials [the 1973 topology delegation] gave me were hand-written, they were records from talks, with many strange symbols I had never seen before, and which could not be found in any books or journals. These were all symbols written down freely by foreign mathematicians during discussions and mutual study, so they could not be found in books or journals, at least not in the short term. So if I wanted to participate in this work, I would have to frequently meet foreign colleagues, frequently go to their seminars and conferences, which put me in a very passive position. So I asked myself, how could I find my own research path, so that I am not subject to foreign influences and can do research even in China?
>
> (Wu Wen-Tsun 2004: 17–18)

The wish to pursue research independently in China was motivated not only by the political insecurity of the late CR period, which made a sustained contact with international mathematics appear unfeasible, but also by Wu's advancing age (he was 54 in 1973). Although he enjoyed the most abstract perspective in his entire career in algebraic topology – he called Sullivan's minimal models a 'beautiful theory' (Wu Wen-Tsun 1987c) – he pursued his highly technical investigations under the self-imposed restrictions of 'calculability' to stay close to 'reality'. This requirement was eventually better satisfied by turning to a completely different field of mechanical proofs, where he became an acknowledged pioneer.

Wu Wen-Tsun's turn to history in the Anti-Lin Anti-Confucius campaign

The Anti-Lin Anti-Confucius campaign, although launched with openly political aims, became an umbrella for historical and philosophical studies. Old books had suffered a lot of damage from the Red Guards during the earlier stages of the Cultural Revolution, but now history was acceptable again.

It is ironic that many Chinese intellectuals found their route to historicism in the midst of a movement that was anything but historicist. The Cultural Revolution was ostensibly motivated by a wish to break the grip of history over the present. But in 1973 and 1974, the radicals themselves mobilized Chinese history as a source of arcane allegories against the perceived opponents of the Cultural Revolution.

This was apparently a coincidence. After Lin Biao's death in 1971, investigation revealed that the dead Marshal – now the chief villain and scapegoat for any social problems – expressed admiration for the ancient sage Confucius (traditionally 551–479 BCE). At the same time, Mao, who had previously made balanced appraisals of the cultural hero, became much more critical of Confucius in a few remarks in 1973. The radicals took up this theme and compared the Cultural Revolution to the First Emperor Qin Shihuang's (259–210 BCE) successful suppression of Confucian opposition to his 'Legalist' policies. Confucians came to represent reaction and the 'Legalists' progressive victorious classes. This interpretative framework was extended *ad absurdum* to all struggles between political and ideological lines, past and present, but first of all to policies and individuals allegedly blocking or reversing the Cultural Revolution, especially the Premier Zhou Enlai (Teiwes and Sun 2007: 118–58, MacFarquhar and Schoenhals 2006: 358–72).

The campaign took various forms in different places. Guan Zhaozhi, who had lost formal power in the factional struggles but still retained part of his ideological authority in IMCAS, suggested studying the history of Chinese mathematics as a respectable contribution to the campaign. Wu Wen-Tsun later explained the appeal of this suggestion:

> Guan Zhaozhi, whom I greatly admired, suggested that everyone should study some ancient mathematics. At that time there was a certain trend of revisiting the past [*fugu* 复古], Guan Zhaozhi's suggestion was relatively legal [*hefa* 合法], because I could not do topology, I was criticized as soon as I started doing it. In that situation, no one had anything else to read, so I also had a look, initially just to know what was there. And I immediately saw what it was like.
>
> (Zhang Zhihui, *et al.* 2008: 105)

IMCAS library records reveal that Wu Wen-Tsun turned in earnest to the history of mathematics in August 1974 (see Chapter 5 for details). Wu's active attitude was praised in the 1974 work report (IMCAS Steering Group 1975a). Wu Wen-Tsun was, as always, quick to show the first results of his studies: on 9 November 1974 he submitted a pseudonymous article to *Acta Mathematica Sinica*, which then appeared in the first issue of 1975 (Gu Jinyong 1975).

Wu alluded by his chosen penname Gu Jinyong to the slogan 'make the past serve the present' (*gu wei jin yong* 古为今用). This was usually meant politically and ideologically, as in this quotation from a more typical article:

> Our study of the history of mathematics must in all cases be permeated by Chairman Mao's principle 'to make the past serve the present', generalize the historical lesson and experience of the struggle between classes and political lines in the history of mathematics using a Marxist standpoint, conception and method, serve the current struggle between classes and political lines, and serve the solidification of the dictatorship of the proletariat.
>
> (Wang Yao 1975: 157)

A similar emphasis was evident in the IMCAS work summary for 1975:

> Last year, we also organized a few people to study history of mathematics. Their work was based on discovering connections between the struggle between Confucians and legalists and the development of mathematics in our country. We want to strengthen even more our research into history of mathematics.
>
> (IMCAS Steering Group 1976)

Wu Wen-Tsun's studies took him elsewhere, however. He wanted to draw lessons from history for current mathematical practice. Instead of the ideological struggle, he elaborated on Mao's policy of 'independence and self-reliance' (*duli zizhu, zi li geng sheng* 独立自主 自力更生), which was also sanctioned in the IMCAS work plan for 1976:

> Uphold the principle of independence and self-reliance, oppose the philosophy of slavishly following the Westerners and prostrating before them, walk our own road of developing science and technology. (...) The labouring masses of workers and peasants of our country have, in the long course of production practice, accumulated plentiful experience. Mathematical inventions and discoveries were not small in number. Some await theoretical summarizing and elevation, providing an important source of development for our mathematical theories and methods.
>
> (IMCAS Party Steering Group 1976)

This could be understood as calling for field study of folk mathematical practices, but Wu extended it to old Chinese mathematical literature. He later claimed that his work on the I^*-functor had benefited from his study of traditional Chinese mathematics, which helped him to overcome the tendency, 'beginning with Euclid in ancient Greece', to separate forms and numbers, always 'intertwined' in traditional Chinese mathematics:

> [T]he emergence and development of algebra was always linked to the development of geometry. We absorbed this thought, and having been inspired by the method of dealing with volumes of solids in the *Shang gong* chapter and elsewhere in the classical mathematical book *Nine Chapters on Mathematical Art*, introduced the concept of 'calculability' of a functor or measure. (...) Ancient Chinese mathematics has a lot we can inherit, elaborate upon and study in order to develop a truly modern socialist mathematics in our country, be it the content and subject matter, the modes of expression, the spirit of its thought or the research methods.
>
> (Wu Wen-Tsun 1976: 15–16)

His inspiration from 'ancient Chinese mathematics' seemed more articulated than the vaguely described contribution of Marxist philosophy:

> These concepts of measure and its calculability are natural and yet completely novel. They are natural, because from the philosophical vantage point of Marxism–Leninism, these are all concepts one will naturally think of. But they are completely new, because before the Cultural Revolution, ignorant as I was about Marxism–Leninism, I would definitely not be able to use these thought weapons so consciously to introduce these concepts.
>
> (Wu Wen-Tsun 1976: 15)

In contrast to drawing on Marxism–Leninism, Wu continued to return to the history of mathematics after the Cultural Revolution and gradually deepened his engagement with the Chinese mathematical tradition. Traditional Chinese mathematics was not guarded by ideological authorities and Wu was thus free to formulate his understanding of its history.

Wu's initial political motivation to study history soon became secondary. At the end of the 1970s, science was rapidly becoming respected and admired. Wu Wen-Tsun was highly regarded for his past achievements. By 1978, he was the only surviving member of the Academic Department of CAS[32] still actively working in IMCAS, and was one of the best-known members of the institute with special treatment:

> Well-known and accomplished scientists such as Chen Jingrun, Yang Le, Zhang Guanghou and Wu Wen-Tsun are all concentrated in one research section of our institute [the Five Subjects Section]. They have many social activities, they accepted about 50 interview requests from newspapers

and other departments. In order to provide them with enough time to do research, and in order to accomplish our organizational work, the business office appointed a person to receive and cater for visitors. In order to have information about the activities of these people and perform our ideological work, the Party branch secretary of this research section, comrade Li Shangjie, personally takes charge of their activities, including presenting their situation, attending seminars, TV and film recordings, proof-reading, even accompanying them to give reports outside the Institute.

(IMCAS Party Steering Group 1978)

But even in this new environment, historical interest remained an important asset. Before 1975, Wu had written only two non-specialist articles for Chinese scholarly journals, both clearly responding to a political command. Wu Wen-Tsun (1956) was his contribution to the launch issue of the *Journal of Dialectics of Nature*, a brief exploration of contradictions in mathematics, especially his familiar topology. Towards the end of the Great Leap Forward, he wrote an elementary survey of important methods in operations research for the *Mathematical Bulletin* (Wu Wen-Tsun 1960b). These articles were both very impersonal and detached from his own work, with no discernible argument. Wu's opinionated post-CR articles on the history of mathematics, on the other hand, helped to establish his new research agenda and demonstrated his identification with China after his intellectual transformation in the 1970s. The 12 years of political campaigns, material and ideological constraints on research, and close encounters with personal tragedies, had apparently only increased Wu's determination to do something for Chinese mathematics. In the absence of standard conditions for mathematical work, he had to mobilize the resources offered to him by his CR experience. In the next chapter, we will see how his historical approach drew on Western and Chinese scholarship, and how it was intimately linked to his mathematical philosophy and new research agenda.

Notes

1 The article was, however, deemed unfit for publication in his selected writings (Wu Wen-Tsun 1986c, 1996).
2 See Dittmer (1991) for a survey of CPC's distancing itself from the CR.
3 For science in general, the event that marked the end of the Cultural Revolution period was the National Conference on Science and Technology in Beijing in March 1978; see, e.g., Yao (1989: 470) or Yuan Zhendong (2008). This broader meaning of the CR period has therefore been adopted in some other recent studies on the history of science in this period, e.g. Schmalzer (2006).
4 A useful survey is in Esherick, *et al.* (2006: 6–16).
5 The story of Chen Jingrun and his sudden rise to national fame will be summarized later in this chapter. See especially Wang Lili and Li Xiaoning (1998); cf. Shen Shihao (1997) and Gu Mainan (2002).

6   'In some ways it will appear as if there were different Cultural Revolutions in different parts of the country' (Esherick, *et al.* 2006: 17).

7   The IMCAS archives hold only payrolls and simple budgets for the years 1967–9.

8   During the Cultural Revolution, someone spiteful changed the original praise of Zheng Zhifu's 'clear class outlook' to 'bourgeois outlook' using a red pencil.

9   There is much debate about the real source of factionalism in this early period. The older, class-based interpretation, elaborated, e.g., in Rosen (1982), seems invalid on many occasions when factional alignment stemmed more directly from responses to current events, such as to the dispatch of work teams to control and suppress the movement (Walder 2002).

10 Only the latter power-seizure is mentioned in the official CAS history (Fan Hongye 2000: 185).

11 Wu Wen-Tsun also took part in a group of younger researchers studying bionics (a discipline that attempts to replicate design from nature in technology) in 1968. Nothing is known about this episode apart from a brief mention in IMCAS (1975).

12 Du Junfu (2009b) claims that this latter phase of the campaign was harsher, and led to suicides in IPCAS too.

13 This happened to Xu Haijin 许海津, a specialist on computational mathematics who had returned from the USA in the late 1950s and had always been in trouble for spending a lot of time abroad (Lu Qikeng 2010; IMCAS 1972b).

14 For more about this movement, see Hu (2005); Qu Jingcheng and Xu Liangying (1984); and Hu Huakai (2006).

15 Compare the insider description of a May 7 Cadre School in Li Yaming (2005: 238–240) with the impressions of an American delegation from a carefully selected May 7 Cadre School in Fitzgerald and MacLane (1977: 90–1).

16 The treatises *On Practice* and *Where does correct human knowledge come from?* were studied collectively at IMCAS in 1971 (IMCAS 1972a). Wu Wen-Tsun was later praised for his thorough knowledge of *On Practice* (IMCAS 1975).

17 The consensus today is that there was never any 'May 16 Conspiracy' in the first place (MacFarquhar and Schoenhals 2006: 221–84). See Du Junfu (2009b) for what it was like to be a suspected conspirator.

18 Zou's suicide was questioned by his widow after the Cultural Revolution (Jing Zhujun 1981).

19 Xu Guozhi, another associate research professor, was also investigated by the leading faction in IMCAS for his previous involvement with the rebels. This factional friction contributed to the split of IMCAS in 1979, after which Guan and Xu headed the newly established Institute of Systems Science.

20 Four of the victims – Xiong Jichang, Zhang Zongsui, Hiong King-Lai and Zou Xiecheng – have already been mentioned. The fourth suicide happened on 1 May 1974, and was motivated partly by political paranoia, partly by chronic disease (IMCAS Steering Group 1974).

21 Two Albanian students had been studying in IMCAS already in 1965–6 (IMCAS 1966b).

22 There has yet been no definitive account of this incident, as many documents remain classified. See Teiwes and Sun (1996), MacFarquhar and Schoenhals (2006: 324–36).

23 Only 20 IMCAS members attended the May 7 Cadre School in 1972 (IMCAS 1973a).

24 The relative fame of Chen Jingrun is of course hard to quantify. A search on the Chinese search engine Baidu returned more than 2.7 million results for Chen, compared to more than 7 million for Hua Loo-Keng, 1 million for S.S. Chern, 950,000 for S.T. Yau, and 900,000 for Wu Wen-Tsun. On Google, Chen with 178,000 came after Wu with 324,000 and Hua with more than 2 million occurrences.

25 The Goldbach conjecture says that any sufficiently large even number is a sum of two primes. Chen's theorem says that any sufficiently large even number is a sum of a prime and a product of at most two primes.
26 The interest in Chen's fate can be linked to a popular romantic stereotype of young, eccentric mathematical geniuses oppressed by the establishment, which has become part of the image of modern mathematics since the nineteenth century (Alexander 2010).
27 On Hua Loo-Keng's popularization movement, see Wang Yuan (1999) and Richard (2008).
28 The most prominent researcher speaking against Deng was actually Chen Jingrun, who wrote a big character poster accusing the rightists of trying to alienate intellectuals from the Party (IMCAS CPC Steering Group 1976a).
29 Thirty-five big character posters were written in the first week of November alone (IMCAS CPC Steering Group 1976f). Significantly, the 'brief reports' were numbered in a new sequence.
30 It was in this quieter period that Wu Wen-Tsun achieved his first results in mechanical theorem-proving (Wu Wen-Tsun 1980).
31 Although Lu claims in this article as well as in other places that the visit occurred in July 1971, and Yang did visit China in 1971, the important visit and talk were more probably in July 1972. Lu mentions 'A talk by C.N. Yang in 1972' as support for his research in IMCAS (1973b), and Zhou Peiyuan wrote his article in that year (Teiwes and Sun 2007).
32 Academic Departments (see Chapter 3) did not convene between 1960 and 1978. The title 'member of the Academic Department' (*xuebu weiyuan*) was never used during the Cultural Revolution.

5 Wu Wen-Tsun's construction of traditional Chinese mathematics

In the standard celebratory account of Wu Wen-Tsun's historical research after 1975, traditional Chinese mathematics is portrayed as a given. Wu simply made correct, penetrating observations about it, and applied them back to his own mathematical research. In the last chapter, however, we have witnessed Wu's increasing efforts to link his research with ideologically legitimate causes. Such ideological concerns became part of his philosophy of mathematics, which he started to articulate by the mid-1970s, and this in turn informed to a large extent his construction of Chinese traditional mathematics.

The word 'construction' does not mean that there had been no traditional Chinese mathematics prior to Wu, or that the qualities he ascribed to it are unfounded fiction. The essential tool of his constructive effort is selection and emphasis, and Wu made no secret that his selection and emphasis was guided by what he considered positive for modern mathematics.

Of course, many of his choices were modelled on previous histories, or reacted to their perceived inadequacies. Wu owed a lot to earlier scholarship and historiography – a fact acknowledged in his work but rarely explained with sufficient clarity. Foreign scholarship was especially important for guiding his attention in both the positive and negative sense. The point of arriving at a better understanding of traditional Chinese mathematics was to recognize Western biases and set the record straight. Wu's writings were quite naturally structured as responses to those Western histories.

But Wu's account was also deeply embedded in Chinese reality and influenced by his Chinese predecessors and colleagues. He drew from the work of the historian Qian Baocong 钱宝琮 (1892–1974), and followed the model of Hua Loo-Keng and Guan Zhaozhi, two mathematicians who made frequent references to the history of mathematics, albeit from very different perspectives. Wu also knew and used the history of Chinese mathematics written by Joseph Needham with the Chinese researcher Wang Ling.

Despite these inspirations, Wu's approach was strikingly original. Let us start the description and analysis of Wu's historicism by a comparison of his first pseudonymous article (Gu Jinyong 1975) with other articles on the history of Chinese mathematics which also appeared during the Anti-Lin Anti-Confucius Campaign (1974–76).

'Making the past serve the present'

Wu came to the history of mathematics during the Anti-Lin Anti-Confucius Campaign, the central thesis of which was the historical struggle between the Confucians and 'Legalists' (*ru fa douzheng* 儒法斗争). The way in which this campaign raged in the history of science has been investigated by Zhang Meifang (2003). Chinese mathematics was integrated into this scheme by identifying authors of Chinese mathematical works who could be more or less plausibly considered 'Legalists'. These typically included the second-century BCE Finance Minister Zhang Cang 张苍, alleged compiler of the *Nine Chapters*, the fifth-century inventor and astronomer Zu Chongzhi 祖冲之, and the eleventh-century polymath and reformer Shen Gua 沈括. 'Confucians' and 'Confucianism' were, on the other hand, blamed for 'idealist and mystical elements' in mathematical work, e.g. the thirteenth-century writer Qin Jiushao 秦九韶, or for oppression and neglect of geniuses such as Zu Chongzhi.

Another recurrent theme was the role of the 'labouring masses' in the production of mathematics, in accordance with Maoist anti-elitism. Here again Confucians played the negative role of suppressing mass creativity and inventing idealist concepts such as 'talent', whereas at least some Legalists could be seen as helping the 'masses' contribute to the development of science and crediting their achievement. The stock example here was Shen Gua's interest in his blind assistant Wei Pu 卫朴, and the fact the he recorded the invention of movable printing type together with its inventor Bi Sheng 毕升 (both Wei Pu and Bi Sheng were presumed to be lowly commoners).

Articles on the history of Chinese science appeared in ideological mouthpieces such as the *Red Flag* (*Hong qi* 红旗) or the *Magazine of the Dialectics of Nature* (*Ziran bianzhengfa zazhi* 自然辩证法杂志) published by Shanghai radicals, but also in scientific journals, recently re-established or newly set up after the Cultural Revolution hiatus. The contents of those published in *Acta Mathematica Sinica* (*Shuxue xuebao*, AMS) and *Mathematics in Practice and Theory* (*Shuxue de shijian yu renshi*, MPT) are summarized in Table 5.1. Among the authors, we find later-distinguished historians of Chinese mathematics (Li Di, Li Jimin) as well as anonymous and pseudonymous teachers and students.[1]

Although it has been suggested that the Anti-Lin Anti-Confucius campaign was designed 'to demonstrate that China had revolutionary power, and revolutionary ideology, before anyone else' (Schram 1989: 181), explicit nationalism was in fact not its main focus, even though current foreign policy issues were part of the conflict between the radicals and the moderates (Teiwes and Sun 2007). Table 5.1 also shows that less than half of the articles deviated from the core ideological issues to praise historical mathematics as a national achievement; when nationalism was present, it was usually linked with 'Legalism'.

Wu Wen-Tsun's take on the theme stood out sharply from the other articles on the history of Chinese mathematics. As can be seen from Table 5.1, his article was the only one that was entirely occupied with the comparative position

Table 5.1 Articles on the history of mathematics published during the Anti-Lin Anti-Confucius campaign

Journal	Title and description	Affiliation
AMS 1974/3	'Notes from studying materials for the history of Chinese mathematics – the struggle during the formation of the *Calendar of Great Clarity* and the mathematical achievements of Zu Chongzhi' (Shu Jin 1974a) Rich historical detail, possibly by an established historian of Chinese mathematics. Praises Zu's innovation against traditional calendar, relatively little ideology, no nationalism.	(uncertain)
MPT 1974/4	'The great progressive scientist Zu Chongzhi' (Shu Qun 1974) Detailed, possibly by someone new to the field. Focus on 'two line struggle': Zu was a Legalist, later Confucians did not understand his books and they were lost. Advanced level of ancient Chinese science as an argument against 'slavish crawling after the West'. Extensive ideological quotations.	(uncertain) IMCAS
AMS 1974/4	'Confucian reactionary thought obstructed and damaged the development of our ancient mathematics' (Shu Jin 1974b) Historical survey of Chinese mathematics, rich but more ideological than the first article by Shu Jin. No nationalism.	(uncertain)
MPT 1975/1	'Shen Gua's mathematical work – supplemented by a discussion of the stimulating role of Legalism on the development of our science and technology' (IMCAS 1974) Rich historical detail, emphasis on Shen Gua's political activities and interest in everyday experience of labouring people. Few ideological quotations, no nationalism.	Operations Research Division IMCAS
AMS 1975/1	'Great contributions of ancient Chinese mathematics to the world culture' (Gu Jinyong 1975) Sophisticated analysis, list of references including Western and Japanese sources, Mao's nationalist quotations, Engels' 'definition' of mathematics from the *Anti-Dühring*.	Wu Wen-Tsun IMCAS

Table 5.1 (cont.)

Journal	Title and description	Affiliation
MPT 1975/2	'A thorough criticism of the absurd theory that mathematics originated from the diagrams of He and Luo' (Li Di 1975) A skilful combination of abundant historical information and derisive denunciations of Confucianism and 'idealism'. Few quotations (only Marx and Engels), list of references.	Inner Mongolia Normal University
AMS 1975/2	'The struggle between Confucians and Legalists and the development of our ancient mathematics' (Beijing Normal Institute 1975) Many ideological quotations, little historical detail, no nationalism.	Students of the Beijing Normal Institute
AMS 1975/3	'About the history of Chinese mathematics' (Wang Yao 1975) Criticism of earlier histories of Chinese mathematics neglecting social and political background. Stresses the role of the masses against isolated 'geniuses', the role of production practice, criticizes mysticism and 'idealism'. Ideological quotations, little historical detail, implicit nationalism (unproductive mathematics inspired by Western missionaries).	Tianjin secondary school
MPT 1975/4	'*Nine Chapters* and the Legalist line' (Beijing Normal University 1975) Focus on the political economy of Western Han as illustrated by practical problems in the *Nine Chapters*. Historically inaccurate, ideological, no nationalism.	Beijing Normal University
AMS 1975/4	'The formation of the *Nine Chapters* and the struggle between the Confucians and the Legalists from pre-Qin to the Western Han' (Li Jimin 1975) Historically rich, heavily ideological argument for an early date of the *Nine Chapters*. No nationalism.	Xi'an Municipal Normal College
MPT 1976/4	'An outstanding medieval mathematician of our country – Li Ye' (Wu Yubin and Chen Xinghua 1976) Some historical detail but mostly ideological, Li Ye as a critic of Confucian thought. Criticism of academic research isolated from social concerns. No nationalism.	Yangzhou Normal Institute

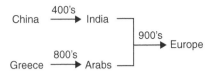

Figure 5.1 The influence of Chinese mathematics on modern mathematics
Source: Adapted from Gu Jinyong (1975: 23).

of Chinese mathematics, the only one not to mention Confucian and Legalist ideology, and the only one to use standard scientific apparatus of references and attributed quotations. He took history much more seriously and much more broadly than the rest of these writers' articles.

The pseudonym Gu Jinyong 顾今用, not intended to hide the author from criticism,[2] was a thinly veiled pun on the slogan 'make the past serve the present' (*gu wei jin yong* 古为今用). The central thesis of the article was the relevance of ancient Chinese mathematics to world mathematics, as expressed already in the title ('Great Contribution of Ancient Chinese Mathematics to World Culture'):

> The development of modern mathematics up to the present day has mainly relied on Chinese mathematics and not on Greek mathematics; it was Chinese mathematics and not Greek mathematics which most importantly decided the direction of historical developments of mathematics.
>
> (Gu Jinyong 1975: 23)

Wu adopted Qian Baocong's idea of two branches of world mathematics merged by medieval Islamic writers ('Arabs'), as indicated in Figure 5.1.

For Wu Wen-Tsun, the alleged Chinese influence was both a question of actual historical transmission (to which his 'answer' was clearly provocative rather than definitive), and a more fundamental question of what is historically significant within mathematics. The relevance of ancient Chinese mathematics was demonstrated by the indispensable role and routine use of the Chinese inventions in modern mathematical activities, whether or not these inventions were actually transmitted to the Hindus, al-Khwarizmi and eventually Descartes.[3]

Throughout the article, Wu hinted that traditional Chinese mathematics conformed remarkably well to a pragmatic philosophy of mathematics, oriented on practical problem-solving in the real world. This was the philosophy of mathematics he had himself adopted by this time, as we will now see.

Wu Wen-Tsun's philosophy of mathematics and mathematical style

Wu Wen-Tsun's philosophical positions can be seen from some of his mathematical and popular writings. His favourite way of doing mathematics also reflects these convictions.

Since the 1970s, Wu has appealed to Friedrich Engels' polemical statement from the *Anti-Dühring* as *the* 'definition' of mathematics:

> [I]t is not at all true that in pure mathematics the mind deals only with its own creations and imaginations. The concepts of number and figure have not been derived from any source other than the world of reality. (...) Pure mathematics deals with the space forms and quantity relations of the real world – that is, with material which is very real indeed. The fact that this material appears in an extremely abstract form can only superficially conceal its origin from the external world.
>
> (Engels [1877–78] 1987: 36–7)

This radical rejection of both mathematical Platonism (the view that mathematical objects exist independently both of human mind and of the real world) and conventionalism (the view that mathematical objects do not really exist, but are freely created by mathematicians) had been orthodox in the People's Republic of China since the 1950s. But Wu only started invoking it after the Cultural Revolution, mostly in the form 'The objects of mathematics are the space forms and quantity relations of the real world.'[4]

This materialist conception of mathematics was prompted by the collective studies of Marx and Engels' works in 1974 (IMCAS Steering Group 1975a). It substantiated the concern for practical applicability of mathematical results and was further reinforced by it when it became internalized. Wu Wen-Tsun had shown this concern, at least rhetorically, during the Great Leap Forward when he wrote about the uses of mathematics in the national economy (Wu Wen-Tsun 1960b), and started to emphasize it from the 1970s:

> Understanding of the logical relationships between concepts has been and always will be important. But one must be aware that if it is pursued too extremely, there is a danger of being caught in inferential play or even number play.
>
> (Wu Wen-Tsun 1987b)

This is an extreme statement with lingering reminiscences of the Maoist era, but even in more recent speeches, Wu has marvelled at the practical applications of seemingly abstract subjects, and emphasized that mathematicians have to be ready to offer their skills to the nation in the event of war or emergency (Wu Wen-Tsun 1999). In a slightly attenuated version, mathematics was practical if it could solve problems. Let us recall that Wu used to be more interested in physics at high school and even during his university studies, and only saw mathematics as a tool. And when being interviewed by Chinese state television in 2006, he elaborated on his pragmatism:

PRESENTER (QU XIANGDONG): Why did you prefer physics [at high school]?
WU WEN-TSUN: I think physics is relatively fundamental [*genben*根本]. If

you want to understand the world, you have to start from physics, not from mathematics. Mathematics is an important tool, but the world can't be understood through mathematics. (…)

PRESENTER: Many people say they feel mathematics is beautiful.

WU WEN-TSUN (LAUGHING): I am not interested in that. I don't have this concept of the 'beauty of mathematics' mentioned so often today. (…)

PRESENTER: So why do you like mathematics?

WU WEN-TSUN: I'm rather a pragmatist. I don't pay attention to whether or not it's beautiful. These are all empty, illusory [*xuwu piaomiao* 虚无缥缈] things. I am not interested in that.

PRESENTER: You like mathematics because it is useful.

WU WEN-TSUN: It is useful. It solves problems.

> (Li Xiangdong and Zhang Tao 2006: 14:30–14:50; Ke Linjuan
> 2009: 149–50)

During the Cultural Revolution, Wu planned to or actually engaged in several cross-disciplinary research projects, including theory of relativity research with Lu Qikeng. Later, he found in problem-solving a correct expression of 'linking theory with practice'. Although he emphasized that this had to be understood broadly and allow for problems that will only appear in the future, a focus on problems was for him the defining feature of good mathematics. This was actually a common view: Wu's teacher S.S. Chern also labelled himself a 'problem-solver' (Jackson and Kotschick 1998), and had castigated Wu for his investigations of logical relationships between topological concepts in 1946 (see Chapter 2).

In order to solve real problems, mathematics should, according to Wu, be constructive. It should provide concrete methods for constructing mathematical objects of desired properties, rather than merely describing their properties abstractly. The constructive mathematical style had in fact been present in much of Wu's work prior to 1975. He often proceeded through series of operations on some imperfect objects to give them properties needed for effective calculations. For example, in his proof of Whitney's product theorem (Wu Wen-Tsun 1948a: 18), Wu first defined a vector field Φ, covering the base complexes of a manifold by linearly independent or zero vectors. Then he transformed it into a vector field Θ, covering the base complexes by linear combinations of constant vectors from the base of the vector space. Finally, he deformed this into a vector field Ψ, covering the base complexes by base vectors multiplied by scalars, which allowed him to directly evaluate the Stiefel characteristic and thus calculate the characteristic class.

A similar line of argument can be seen in Wu's work on imbedding (Wu Wen-Tsun 1955d). He noted that the key feature of imbedding mappings is their injectivity (no two distinct points are mapped into the same point when imbedded). But homotopy-type methods of algebraic topology do not preserve this property. To 'highlight the injectivity of imbedding, and only then use methods of homotopy' (Wu Wen-Tsun 1965a: ix), he constructed from a given complex K a space of all ordered pairs of its points $\{x, y\}$ $\tilde{K}^*$, and from this again a space of all unordered sets $[x, y]$ K^*. Readily calculable invariants of K^* – cohomology classes – were not homotopy invariant for the original space K. In a series of constructions, Wu imbedded lower-dimensional skeletons of the complex K (i.e. all vortices, all 1-simplexes, etc.) in corresponding Euclidean spaces:

> Let $\mathbf{R}^1 \subset \mathbf{R}^2 \subset \ldots \subset \mathbf{R}^m$ be a sequence of linear subspaces of increasing dimension in $\mathbf{R}^m$. By trying to realize the complex K in a certain canonical manner such that $K^0 \subset \mathbf{R}^1$, $K^1 \subset \mathbf{R}^3$, etc., representative cocycles Φ^m [from the product space K^*] may be explicitly constructed. This gives us not only the means to compute effectively these classes in every concrete case, but also makes it possible to derive a series of properties of Φ^m which are not easy to foresee.
>
> (Wu Wen-Tsun 1955d, with grammar corrected)

Similar, more complicated, examples may be found in Wu's other works. Generally speaking, his style of argument has almost always been constructive in a loose sense, relying on subsequent constructions or on the choice of suitable objects. His results are often elaborate summaries and generalizations of his experience with mathematical objects, rather than a series of logical deductions from their properties. Arguments based on logical consistency are marginal in his writings.

Wu fully realized the importance of constructive arguments for applications in the 1970s when he was adapting his theory of immersion to the design of integrated circuits. He also emphasized the 'calculability' of the 'I^*-functor' (Sullivan minimal models) which he explored in the 1970s. Later he formulated his position more clearly and set it into the context of the development of mathematics by advocating a 'revival of constructive mathematics':

> [A]pplications acutely need constructive mathematics. (...) There are many problems in practical research which it can be difficult to solve constructively for some time, therefore existence, possibility and related problems are studied first, but eventually [mathematics] should be constructive. It is worth noting that, due to stimulation by various factors, the constructive outlook has recently clearly had an ascending tendency.
>
> (Wu Wen-Tsun 1985a: 334)

Wu also adopted a constructive approach in his book Wu Wen-Tsun (1987c), and later expressed his agreement with Leopold Kronecker (1823–91), one of the best-known proponents of constructivism in modern mathematics, against David Hilbert (1862–1943) and his famous non-constructive proof of the finite basis theorem (Wu Wen-Tsun 2000: iii).

An important point for understanding Wu Wen-Tsun's historical writings is his attitude to axiomatization, a tendency in modern mathematics which originates in the development of geometry. After the original system of axioms in Euclid's *Elements* was modified to accommodate various non-Euclidean geometries in the nineteenth century,[5] David Hilbert created in his famous *Foundations of Geometry* (1899), an abstract axiomatic system which can describe the objects of these various geometries as well as 'tables, chairs and beer mugs' (Reid 1970: 208).[6]

From then on, 'axiomatization' became an increasingly common method of providing relatively logically secure foundations for various mathematical theories. Relative logical security was achieved by reducing theories into statements about sets of primitive objects, which were defined only by appearing in the axioms, rather than by their intrinsic properties. Wu Wen-Tsun was fully exposed to this method in France, where it was used extensively by Nicolas Bourbaki.[7]

Nicolas Bourbaki is the pen name of a group of mostly French mathematicians, formed in 1934–35, with the initial aim of writing a modern textbook of mathematical analysis, and later a comprehensive, unifying account of modern mathematics. The group has existed around the publication project ever since, although it became less active since the 1970s. Their serial publication, the *Elements of Mathematics* (*Éléments de mathématique* – the departure from the usual French plural form *mathématiques* signals the perceived unity of mathematics), consists of expositions (*exposé*) of fundamental branches of mathematics (set theory, algebra, topology, function theory, etc.), drafted and edited at collective meetings. Wu Wen-Tsun's French teachers Charles Ehresmann and Henri Cartan were among the founders of Bourbaki, and S.S. Chern also joined their meetings during his stay in Paris in 1937–38, although not those at which *Elements of Mathematics* were drafted (Tian Miao 2000; Jackson and Kotschick 1998). In 1948, the Bourbaki mathematicians established a regular public seminar (*Séminaire Bourbaki*) to brief each other about recent world developments in mathematics. When Wu Wen-Tsun came to Paris in 1949, he also attended its sessions.

Bourbaki tried to unify mathematics around the notion of structures – sets equipped with axioms describing the possible relationships between their elements. Their project was one of the highest points of the tendency towards axiomatization. However, rapidly developing branches of mathematics did not easily yield to such a constraining approach. This was the case in algebraic topology, in which Wu Wen-Tsun worked. Indeed, Wu Wen-Tsun's work reveals little interest in axiomatization. On the contrary, he took special delight in his ability to prove the equivalence of the Smith calculus of periodic transformations with Steenrod squares intrinsically, using his extensive experience with the Smith calculus in proofs of imbedding theorems, rather than with the help of Thom's axiomatization of Steenrod squares (Wu Wen-Tsun 1957c).

Axiomatization was an ideologically contested approach in China. Guan Zhaozhi had initially welcomed it as a new stage of mathematical abstraction, which would make mathematics applicable to more diverse ends (Guan

Zhaozhi 1957). After 1957, however, Guan joined those who warned against axiomatization as obscuring the ultimate origin of mathematical knowledge in production practice (Guan Zhaozhi 1959). Axiomatization was also blamed for misleading young mathematicians into an 'idealist academic philosophy' (Jiang Zehan, *et al.* 1958). The most thorough criticism of axiomatization occurred during the CR period:

> Bourgeois idealists have promoted the axiomatic system of Euclidean geometry and the birth of Lobachevsky's geometry as a trump-card for the view that 'the mind creates mathematics', and vehemently denied that geometric axioms come from practice, deceptively claiming that mathematical axioms are heaven-bestowed truths, and raising axiomatic method into supreme position. (...) Some people wasted their whole life influenced by this philosophical view, pointlessly wanting to create from nothing axioms of their own and establish themselves as authorities in the history of mathematics, to spread their name across the world.
>
> (Jinhua Teachers College 1972: 46)

Others subtly defended axiomatization (Sun Yongsheng 1972), and Wu Wen-Tsun explored axiomatization in various contexts after the 1970s. He gave a course about the axiomatization of geometry in 1978–79, and investigated implications of Hilbert's *Foundations of Geometry* for mechanical proofs (Wu Wen-Tsun 1982b). His first book on mechanical theorem-proving (Wu Wen Tsun 1984b) also systematically studied the axiomatic foundations for the mechanization of geometry. But at the same time, he came up with a concept of 'mechanization' as an alternative to axiomatization, which to him seemed to have gone too far in modern mathematics (Wu Wen-Tsun 1981c). The lack of axiomatic systems in traditional Chinese mathematics became an intriguing feature and possible asset for him. This was precisely opposite to the assessment of Western histories of mathematics, which he studied quite closely at the time.

Wu Wen-Tsun's reaction to Western historiography

According to Wu's later recollections, the only knowledge of the history of mathematics he had before 1974 came from Western histories. He had bought some books about ancient Chinese mathematics but never managed to read them before they were lost or destroyed in the early phase of the Cultural Revolution (Li Xiangdong and Zhang Tao 2006).[8] In 1974, he delved into historical literature, part of which was borrowed from the IMCAS library. The paper slips in those books still recorded his name and borrowing dates (19 August and 19 October 1974) in 2008. His first article (Gu Jinyong 1975), however, only quoted Cajori and Scott, and listed many additional sources (Table 5.2).

Table 5.2 Wu Wen-Tsun's library borrowings and listed sources of Gu Jinyong (1975)

Source	Note
Cantor, M.: *Vorlesungen über Geschichte der Mathematik.* Teubner, Leipzig 1894	Borrowed, not quoted
Cajori, F.: *A History of Elementary Mathematics, with Hints on Methods of Teaching.* Macmillan 1917	Borrowed, quoted
D'Ocagne, M.: *Histoire abrégée des sciences mathématiques.* Vuibert, Paris 1955	Borrowed, not quoted
Scott, J. F.: *A History of Mathematics.* Taylor and Francis, 1958	Borrowed, quoted
Mikami, Y.: *The Development of Mathematics in China and Japan.* Teubner, Leipzig 1913	Not borrowed, quoted
Smith, D. E.: *History of Mathematics.* Ginn & Co., Boston 1925	Not borrowed, quoted
Datta, B., Singh, A.N.: *History of Hindu Mathematics – A Sourcebook.* Motilal Banarsi Das, Lahore 1962	Not borrowed, quoted
Struik, D. J.: *A Concise History of Mathematics.* Dover, New York 1948	Not borrowed, quoted
Bourbaki, N.: *Éléments d'histoire des mathématiques.* Hermann, Paris 1969	Not borrowed, quoted
Qian Baocong: *Zhongguo shuxue shi* [History of Chinese Mathematics]. Kexue chubanshe, Beijing 1964	Not borrowed, quoted
Needham, J.: *Science and Civilisation in China*, vol. 3 (Mathematics), 1971 [a reprint, original publication 1959]	Not borrowed, quoted

European and American books formed Wu Wen-Tsun's initial, quite dismissive, view of Chinese mathematics. He had considered Chinese mathematics to be just simple arithmetic, without anything worth his attention (Li Xiangdong and Zhang Tao 2006). This is hardly surprising in light of the cursory and unconvincing treatment Chinese mathematics received in some of his readings:

> A rudimentary form of land measurement, as with other races of antiquity, led to a knowledge of certain geometrical forms. (…) The eleventh and twelfth centuries were barren. There was a little activity during the thirteenth; after that there was but slight development in mathematics on Chinese soil until more recent times. (…) Valuable as the contributions of the Chinese were, particularly at a period when mathematical progress was retarded, the subject was no more abstract in their hands than it was with the Hindus.
>
> (Scott 1958: 80–2)

Scott devotes less than three pages to the entire history of Chinese mathematics, compared to 14 on Indian mathematics. Even less on China could be found in the edition of the influential *Concise History of Mathematics* available to Wu (Struik 1948). This book had been translated into Chinese by Guan Xian 关娴 (Guan Zhaozhi's sister) in 1956, with a long preface by Guan Zhaozhi, which was also published separately as a critical review (Guan Zhaozhi 1956b).[9] Wu Wen-Tsun, however, referred to the first English edition. Struik had devoted six pages together (in a smaller format than Scott) to both Chinese and Hindu mathematics, and passed a very unkind verdict on the entire 'Oriental mathematics':

> Nowhere in all ancient Oriental mathematics do we find any attempt at what we call a demonstration. No argumentation was presented, but only prescription of certain rules: 'Do such, do so'. (...) To us who have been educated on Euclid's strict argumentation, this whole Oriental way of reasoning seems at first strange and highly unsatisfactory. But this strangeness wears off when we realize that most of the mathematics we teach our present day engineers and technicians is still of the 'Do such, do so' type, without much attempt at rigorous demonstration. Algebra is still being taught in many high schools as a set of rules rather than a science of deduction.
>
> (Struik 1948: 31–2)

This passage, dismissive as it was, had a direct and lasting influence on Wu. In a curious case of 'transvaluation of values', he referred to it repeatedly as a proof of the relevance of Oriental – or more precisely Chinese – mathematics in the modern world, as well as of the Western bias and Eurocentrism:

> What this book (i.e. Struik 1948) calls Oriental mathematics originally refers to Babylon or India, whereas in fact it should be referring to China to be correct. The branches of arithmetic and algebra in elementary and middle schools, from number notation to solution of systems of quadratic equations, are essentially inventions and creations of ancient Chinese mathematicians.
>
> (Gu Jinyong 1975)

Wu Wen-Tsun's approach to the history of Chinese mathematics continued to be formed by interactions with Western scholarship after his initial article had been written. IMCAS library records show that in 1976 he went back to Smith's and Scott's histories between February and June, and kept Heath's translation of Euclid for almost the entire year. Also in 1976, Wu borrowed a German-language history of mathematics (Hoffmann 1953–57), which did not appear as reference in any of his writings, but possibly guided him to an earlier German historian, H. Hankel, quoted in Wu Wen-Tsun (1987b). Wu's last recorded borrowings in the IMCAS library were in 1977 (Scott, Smith, Heath and Cajori).

These mainstream Western histories were used for two slightly contradictory purposes. On one hand, Wu criticized their methodological biases with regard to China and responded to their deficiencies. On the other hand, he relied on them for accounts of rival traditions – Indian, Babylonian and Greek. This mixture is evident in Gu Jinyong (1975); its further expression was in Wu Wen-Tsun (1982c), probably the most impressive of Wu's historical texts. Here Wu evaluated ancient Chinese surveying and geometry as an example of the special character of Chinese mathematics. His aim was to reconstruct this specific character by avoiding Western-style explanatory devices not attested in China, including reasoning with parallel lines, and with modern algebraic symbols. He criticized the practice common in earlier histories of mathematics – Struik (1948) and Scott (1958) – to draw wide-ranging connections between earlier and later mathematics. Mistaken reconstructions of ancient Chinese surveying methods were for him a symptom of 'a mistaken methodology which has already spread to most widely circulating histories of mathematics and not only obscured the true facts about ancient mathematics, but distorted it beyond all recognition, being a major reason for the emergence of many myths about Babylon, India and Diophantos' (Wu Wen-Tsun 1982c: 19).

Wu singled out Struik's and Scott's claims about Babylonian mathematics as examples of this problem. They were exaggerating that 'the Babylonians of Hammurabi's days were in full possession of the technique of handling quadratic equations' (Struik 1948: 26) on the basis of fragments of Babylonian numerical calculations. Apart from the overt concern about methodology, the main source of Wu's uneasiness was the priority question. He went on to explain that knowledge only grows through slow and gradual accumulation, and showed for the Chinese case that later techniques cannot be attributed to earlier forerunners:

> We could say that the geometric solution methods in the *Explanation and diagrams of Base, Altitude, Circle and Square* [*Gou gu fang yuan tu shuo* 句股方圆图说, a third century text by Zhao Shuang 赵爽] and the *Nine Chapters* are a source and origin of our algebra of the Song and Yuan periods, having a certain promotional and inspirational effect, but we definitely can not regard them as algebra. For the same reason, although we respect great achievements of ancient Babylonian people, to carelessly attribute modern algebra to them is, it must be said, extremely absurd.
>
> (Wu Wen-Tsun 1982a: 20)

Wu came up with the rhetorically effective label 'mechanization' for both his new method of automatic theorem-proving and for traditional Chinese mathematics (Wu Wen-Tsun 1978e, 1980). But despite this rhetorical effectiveness, the label, and especially the axiomatization–mechanization dichotomy, was not quite satisfactory from the historical perspective. There were no calculating machines in ancient China or other Oriental civilizations – why should their mathematics be more 'mechanical' than the Greeks'? Moreover, although Greek geometry was axiomatized, axiomatization was not its working method in the same way as in modern mathematics. Perhaps because of these problems, in his later works Wu introduced an opposition between proofs and algorithms as a replacement. A crucial influence in this shift was the work of the computer scientist Donald Knuth. Wu studied Knuth's textbook *The Art of Computer Programming* (first volume 1968, second 1969, third 1973), which consists of commented algorithms, just like ancient Chinese mathematical classics consisting of problem-solving methods. Knuth also wrote an article on ancient Babylonian algorithms (Knuth 1972). Although never cited by Wu himself, it was mentioned by his younger colleagues Li Wenlin and Yuan Xiandong (1982) in the same volume as Wu Wen-Tsun (1982c). Both Li Wenlin and Karine Chemla, who studied ancient Chinese mathematics in Beijing in the early 1980s, confirmed the influence of this article on Wu Wen-Tsun's thought (personal communication).

Wu might as well have drawn inspiration from Knuth's opening sentence: 'One of the ways to help make computer science respectable is to show that it is deeply rooted in history, not just a short-lived phenomenon.' Wu Wen-Tsun had a similar motivation for his own turn to history, although it could also be said that he proceeded in the opposite direction, making ancient Chinese mathematics respectable by showing what can be rooted in it.

Knuth drew a series of analogies between ancient Babylonian mathematics and computer science: Babylonian sexagesimal notation was actually the first floating-point notation; their algebraic algorithms were 'machine language' as opposed to the 'symbolic language' of our modern algebra; they used numerical algebra disregarding physical and geometrical significance. Knuth also compared particular algorithms to a 'stack machine' or to a 'macro expansion'. His article was not a serious history of algorithms, but rather a reminder of the venerable ancestry of the basic techniques of computer science. But Knuth's last paragraph must have been very suggestive to Wu:

> What about other developments? The Egyptians were not bad at mathematics, and archaeologists have dug up some old papyri that are almost as old as the Babylonian tablets. The Egyptian method of multiplication, based essentially in the binary number system (...) is especially interesting. Then came the Greeks, with an emphasis on geometry but also such things as Euclid's algorithm; the latter is the oldest nontrivial algorithm which is still important to computer programmers. (...) And then there are the Indians, and the Chinese; it is clear that much more can be told.
>
> (Knuth 1972: 676)

In comparison to Knuth's article, Wu's sole emphasis on Chinese mathematics appears narrow-minded. It might even be suggested that Wu tried to take away some credit from 'rival' ancient civilizations in his later attempts to demonstrate the computational superiority of specific Chinese algorithms over Western ones. He devoted considerable attention precisely to Euclid's famous algorithm for the greatest common divisor of two numbers:

Propositions 1 and 2 of book 7 of the Elements are indeed equivalent to what later generations called Euclid's algorithm, but given the original formulation of their proofs, they are neither constructive nor mechanical. Even after reformulation by later mathematicians into the form of a true algorithm, it is only an exception in the entire Euclidean system. It cannot be discussed in the same breath with constructive and mechanical algorithmic results permeating our ancient mathematics.

(Wu Wen-Tsun 1987a: 76)

Wu Wen-Tsun also greatly admired the work of two Western historians of Chinese mathematics: Ulrich Libbrecht and Donald Wagner. Libbrecht's study of Qin Jiushao's *Mathematical Treatise in Nine Chapters* (*Shu shu jiu zhang* 数书九章, 1247) was mentioned repeatedly by Wu as an example of valuable scholarship, which should shame Chinese mathematicians incapable of understanding their tradition in the same depth. The two men met in Beijing in the early 1980s, and Wu drew from the encounter a new argument for his mistrust of Western histories of mathematics:

A few years ago, a Belgian historian of mathematics called Libbrecht came to China to give talks,[10] which I also attended. He said something to this effect: Greece has been paraded so high because Europeans at the time wanted to recover from a backward state and needed some ideology to raise their morale. He used the word 'Eurocentrism' (which I heard for the first time), and it means that all culture originates in Europe, you in the East are all savages.

(Wu Wen-Tsun 1984c)

In the same speech, Wu also expressed his admiration for Needham and his linguistic skills, but they never met. Neither did Wu ever meet D. Wagner, but he has praised him highly for his two articles discussing the Liu Hui commentary on the *Nine Chapters* (Wagner 1978, 1979). Wagner's second paper was translated into Chinese in 1980, and Wu discovered it by about 1986. He was thrilled by the fact that Wagner had demonstrated complex logical inferences in Liu Hui's commentary, never observed by earlier readers.[11]

In 1982, Wu once more visited Paris and participated in one of the Bourbaki seminars again. This inspired him to organize a similar seminar in Beijing in October 1985, called the *Liu Hui Seminar*, after the third-century commentator to the *Nine Chapters*. Fifteen outstanding Chinese mathematicians presented review reports on recent developments in their disciplines, which were then published as *New Advances in Modern Mathematics* (Wu Wen-Tsun 1988c). In the preface to this volume, Wu explained his inspiration in the Bourbaki seminars, and evaluated Bourbaki as follows:

Since the 1950s, the influence of Bourbaki has already inundated the entire mathematical community. Young mathematicians have in great numbers taken Bourbaki as their model and the *Elements of Mathematics* as the foundation of their study (...). These are all great achievements of the Bourbaki school, but they are only an outward expression, not sufficient to explain their spiritual substance.

The author has met a Third World mathematician abroad who said: 'Bourbaki is the product of the French national spirit.' This is indeed an incisive comment. The real Bourbaki is the Bourbaki he spoke about. (...)

Recently, the influence of Bourbaki has already started to wane (...). But the effort they have made to reinvigorate the spirit of France should not only be admired by the French people, but also studied and applied by other nations. What we should learn from the Bourbaki school is not their specific achievements in various branches, nor their often controversial philosophical system. These can but do not have to be learned and followed. What we truly should learn from them is this admirable spirit.

(Wu Wen-Tsun 1988c, quoted from Wu Wen-Tsun 1996: 46–8)

Wu's creative work with the history of mathematics was thus inspired by the 'national spirit' and 'the truly admirable' perseverance of the French mathematicians in the service of their nation. This does not sound like a fair characterization of Bourbaki's main concerns on Wu's part, but it shows brilliantly his appropriation of foreign inspirations to construct a rhetorically effective history. Most of the other examples in this section were used by him in a similarly instrumental way.

The last foreign influence to merit a brief mention is the Indian mathematician Shreeram Abhyankar (born 1930), a distinguished algebraic geometer. In 1976, he published an article appropriately entitled 'Historical ramblings in algebraic geometry and related algebra', in which he distinguished high-school algebra (algebraic algorithms), college algebra (group theory, rings and modules, etc.) and university algebra (the use of functors and categories). He claimed that his thorough knowledge of 'high-school' algebra was the decisive factor in his successes in algebraic geometry, whereas 'college' and 'university' algebras often hindered research by subsuming only a few of the useful techniques of earlier periods. Abhyankar also traced the origins of 'high-school' algebra to ancient Indian mathematicians. He had learned as a boy the Sanskrit treatise *Siddhanta shiromani* by the twelfth-century Bhaskara II, and used the techniques he memorized there in his research (Abhyankar 1976).

Abhyankar's article received two awards from the Mathematical Association of America, a body with a strong interest in the didactics of mathematics and thus also 'high-school algebra'. The Indian connection was no doubt regarded – and possibly intended – as a personal curiosity, related to the author's Brahmin descent. His praise for 'high-school' algebra was not limited to Hindu mathematicians, but directed largely at classical mathematicians from Newton to Macaulay. Wu Wen-Tsun followed a similar path in his advocacy for mechanization, which was not only present in Chinese mathematics, but also in the work of early modern European thinkers such as Descartes and Leibniz.[12]

Chinese priorities and mathematical style

When writing about the misrepresentation of traditional Chinese surveying (Wu Wen-Tsun 1982c), Wu was reacting not only to Western historians, but to Chinese and Japanese ones as well. He protested against the usual explanation of how the method of double differences was established. The method has been used since the third century AD at the latest to calculate distances, heights and depths. A target above or below the observer was spotted from two or more points at known distances, while measures were taken of the vertical deviation (i.e. the tangent) of the line connecting the target and the observer's eye. The simplest example is the calculation of the height of the sun (Figure 5.2).

Under the assumption of a flat earth, the height of the sun is given by the formula $H = g(d_2-d_1)/(y_2-y_1) - g$. In many earlier explanations of this formula, historians such as Yoshio Mikami (1875–1950) and Qian Baocong used parallel lines, proportional calculus or algebraic manipulations for demonstration. In Wu's view, this distorted the achievements of ancient Chinese surveyors: they appeared to be deriving their methods by non-systematic, intuitive application of principles which were only truly developed in medieval or modern Western mathematics. Instead, Wu reconstructed a small set of simple geometric principles which were used consistently in many Chinese treatises throughout history. Chief among them was the so-called 'Out–In Complementary Principle' (*chu ru xiang bu yuanli* 出入相补原理, OICP), stating (in Wu's words) that 'a planar figure does not change its area when moved from one place to another' (Wu Wen-Tsun 1978a: 81). This general principle enabled breaking up complex shapes and evaluating the area of their components, creating a simple but powerful graphical calculus. Wu argued that the 'Sun Height Method' was derived by Chinese mathematicians Zhao Shuang and Liu Hui (both third century AD) by manipulating diagrams, moving around and comparing simple rectangles. The basic application of OICP to the Sun Height Diagram is a proof that corresponding rectangles on both sides of the diagonal are of equal area. (The correspondence is indicated in Figure 5.2 by the inclination of hatch lines. The chequered bits are where both hatches are superimposed.) From this it follows that gd_1 equals $(H-g)y_1$ and gd_2 equals $(H-g)y_2$. By invoking OICP, their differences must also be equal, thus $g(d_2-d_1) = (H-g)(y_2-y_1)$. Finally, the rule of three is used to arrive at the height sought.

Wu emphasized the pervasiveness of OICP in Chinese mathematics (Wu Wen-Tsun 1978a, 1981a, 1982a) and the error of explaining the ancient methods by essentially Greek demonstrations in order to show that Chinese geometry, far from being only an imperfect version of Western geometry, was

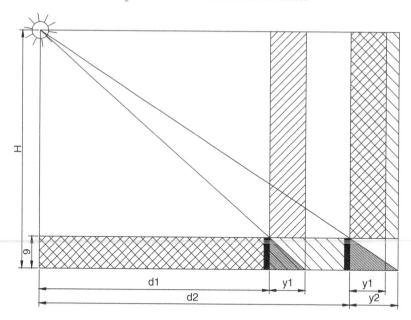

Figure 5.2 The Sun height diagram
y_1, y_2: gnomon shadow lengths
g: height of gnomons
d_1, d_2: distances of gnomons from the point 'under the sun'
H: Sun height
Source: Adapted from Wu Wen-Tsun (1982a: 196).

advanced and systematic in its own way. Use of modern and Western explanatory tools obscured this specific character and sophistication, which Wu
showed by an analogy with ascribing knowledge of calculus to Archimedes:

> It causes the glorious ancient achievements to vanish without trace. This
> is just as when post-seventeenth century calculus, a modern weapon, is
> used to prove Archimedes' formula about the area of a parabolic arc.
> This is of course very easy today, it is a simple exercise even for those who
> have just studied a little calculus. But if we use [calculus] to demonstrate
> that Archimedes' formula and theorem were correct, it is not only com-
> pletely insignificant, it even turns history upside down and devalues the
> major contribution of Archimedes.
>
> (Wu Wen-Tsun 1982c: 18)

Wu criticized the type of reconstructed proofs suggested by Qian Baocong
(without explicitly naming the already deceased historian) as 'not only not in
accordance with the true nature of ancient Chinese geometry, but somewhat
more harshly put (...) "wrong"'. But despite this criticism, Wu acknowledged

Qian as a great historian of mathematics and relied heavily on his book (Qian Baocong 1964) for various purposes. He drew his generalizations and conclusions from Qian, but it was also important for highlighting to Wu a Chinese algebraic technique from the fourteenth century, which he built into his method of mechanization, as will be discussed in the next chapter.

But Wu first of all followed Qian in his speculation about connections between Chinese mathematics and the mathematical traditions of other great cultures, a line of argument first attested in Qian's 1951 address to secondary school teachers of mathematics. In this article, Qian Baocong listed the strengths of Chinese mathematics in comparison to Greek mathematics, and concluded his exposition with words that were quoted at length at the end of Wu Wen-Tsun's pseudonymous first historical article (Gu Jinyong 1975):

> After the 5th century, most of Indian mathematics was of the Chinese style, after the 9th century, most of Arabic mathematics was of the Greek style, in the 10th century these two schools of mathematics merged and propagated, through the Muslims of Northern Africa and Spain, to all parts of Europe, and thus Europeans on the one hand recovered Greek mathematics they had already lost, on the other hand they absorbed high-vitality Chinese mathematics, and only then did modern mathematics start to develop dialectically.
>
> (Qian Baocong 1951: 1043)

Qian's interest in the question of influences resulted in a compilation of lists of suspected mathematical transmissions from China to India and Muslim countries. A similar idea occurred to Joseph Needham and Wang Ling, who included a famous 'quite considerable list' of 14 mathematical ideas that 'radiated from China southward and westward' (Needham and Wang Ling 1959: 146), including decimal notation, negative numbers and unified treatment of algebraic and geometric problems. This list was considered one of the best parts of Needham and Wang Ling's book by its otherwise critical Chinese reviewers (Du Shiran and Mei Rongzhao 1982).[13] Such lists of priorities were compiled and studied as a way of evaluating the relevance of Chinese mathematics. Qian Baocong, Needham and Wang Ling, and Wu Wen-Tsun all had distinct focuses, but their lists also share many items (Table 5.3).

Needham and Wang Ling wanted to highlight items which were both Chinese priorities in the global context, and plausible objects of transmission. Qian Baocong was interested in ideas and results which where attested earlier in China than India and thus possibly transmitted to Hindu mathematics, even if they had been known in the Greek world even before that. Wu Wen-Tsun's table, finally, focused on the development of Chinese arithmetic and algebraic computational algorithms. Like Needham, he chose only those techniques which he believed Chinese mathematicians created first in the world, but he did not constrain his interest to cases of demonstrated (or plausible) transmission. Thus Chinese skill with equations, especially of

Table 5.3 Chinese priorities in Needham and Wang Ling, Qian Baocong, and Wu Wen-Tsun

Order of listing in	(Needham 1959: 146)	(Qian Baocong 1964: 109–12)	(Gu Jinyong 1975: 19)
Decimal place-value notation	1	1	1
Basic arithmetic operations	—	2	—
Square and cube roots	2	13	4
Rule of Three	3	4	5 (Applications)
Fractions	4	3	2
Decimal fractions	—	—	3
Negative numbers	5	7	6
Proof by diagram of the Pythagorean theorem, Broken Bamboo Problem[1]	6–7	8	—
Area of a circle segment	8	5	—
Identity of algebraic and geometrical relations	9	—	[p. 21]
Rule of false position	10	[p. 220]	—
Indeterminate analysis, The Hundred Fowls Problem[2]	11–12	11–12	—
Systems of linear equations	—	6	7
Quadratic equations	—	—	8
Cubic numerical equations	13	—	9
Higher-degree equations	—	[roots p. 220]	10
Systems of higher equations	—	—	11
Pascal triangle	14	—	—
Volume of a sphere	—	5	—
$\pi \cong 3927/1250$	—	9	—
Double difference	—	10	—
Halving for calculation of sines	—	14	—

Notes:
[1] The Broken Bamboo Problem is to find, using the Pythagorean theorem, the height at which a bamboo has to be broken so that its tip reaches a specified distance from the root. This is problem 9.11 of the *Nine Chapters on Mathematical Art*.

[2] The Hundred Fowls Problem is to find, by indeterminate analysis, the number of roosters, hens and chicks bought for 100 coins given their respective prices. This is problem 3.38 in the fifth-century *Mathematical Classic of Zhang Qiujian* (Lam Lay Yong 1997: 235).

higher degrees, is also acknowledged in both Needham's and Qian Baocong's books, but given that this Chinese art did not exercise observable influence on Indian or Western mathematics, it does not appear in those lists tracing transmissions.

Wu Wen-Tsun's list was tightly focused on the development of essentially algebraic techniques, and on the connection between Chinese geometry and algebra. This connection was a prominent characteristic for Needham and Wang Ling as well, who led Wu Wen-Tsun to realize that 'unified treatment of geometry and algebra was a traditional feature' of Chinese mathematics (Gu Jinyong 1975: 21). Wu Wen-Tsun considered this 'unified treatment' to be a central component of the Chinese mathematical style, and one especially worth preserving, as he indicated in *The Great Chinese Encyclopedia*:

> In the real world, numbers are to shapes as their shadows, completely inseparable. Ancient Chinese mathematics reflected this objective fact, numbers and shapes have always supported each other and generated each other, developing in parallel.
>
> (Wu Wen-Tsun 1988a)

Wu Wen-Tsun's appeals to this tendency in ancient Chinese mathematics go back to 1976, when he argued that it guided his work on the theory of minimal models (I^*-functor):

> Western mathematics follows the tradition started by Euclid in Ancient Greece and often separates shapes and numbers, either becoming too geometrical and intuitive, or too abstract in the form of pure algebra. Mutual communication of shapes and numbers and their unified treatment often are not pursued actively, or [shapes and numbers] are even consciously separated. In contrast, our ancient mathematics had in its development always blended spatial forms and quantitative relations together, and the emergence and development of algebra was therefore always intertwined with the development of geometry. We have adopted this thinking.
>
> (Wu Wen-Tsun 1976: 15)

Wu's understanding of Chinese mathematics was also founded upon the ideas of Li Jimin 李继闵 (1938–93), whom he greatly praised. This gifted mathematician, originally an expert on function theory, was cut off from research while sent down to rural areas of north-west China from 1962 to 1974. In the early 1970s, he was able to resume a position in a department of mathematics in Xi'an, but opted for the history of Chinese mathematics instead of trying to renew his early career as a research mathematician. Li Jimin's historical papers tried to discover within Chinese mathematical algorithms traces of their supporting theoretical structures, which were never explicitly recorded in Chinese mathematical literature. Li described Chinese mathematics as 'hiding the theory [or principle, *li*] in calculation' (*yu li yu suan* 寓理于算, an allusion to the long Chinese tradition of 'hiding history in biography' *yu shi yu zhuan* 寓史于传), pointing out that 'in the past, the algorithmic expressions of traditional Chinese mathematics were misunderstood as lacking theory, but in fact, any systematic algorithm is guided by a certain theory' (Li Jimin 1987: 250); cf. (Li Jimin 1984: 13–14). Wu found this way of looking at algorithms very helpful and frequently alluded to it (Wu Wen-Tsun 1987a). In 2010, Wu emphasized that 'many of my opinions about ancient Chinese mathematics came from [Li Jimin]' (interview with the author).

Chinese mathematicians' use of the history of mathematics

Wu Wen-Tsun had not written on the history of mathematics before 1975. So far, we have seen how his decision to enter this field was motivated by contemporary politics as well as long-term personal interests. But Wu was also aware of the use others around him had made of the history of mathematics for the promotion of their mathematical ideas. This was an important background against which his writings should be read.

The two main leaders of IMCAS, Hua Loo-Keng and Guan Zhaozhi, had both referred to the history of mathematics before the 1970s. They chose different historical facts and put the history to different use, but both shaped the climate in which it was respectable to insert historical material into mathematical and polemical literature. Wu Wen-Tsun's historical writing provides a sharp contrast to their style, but was also enabled by their opening up of this type of argument.

Hua Loo-Keng was very fond of Chinese traditional mathematics, and portrayed it as an almost folk phenomenon. He was also a skilled abacus calculator, according to Lin Qun (personal communication 2008). Shortly after his return to China, he published an article in the *People's Daily* called 'Mathematics is a discipline in which the Chinese excel'. It was designed as an antidote to the invasion of Western cultural imperialism in preceding decades, which made people 'forget one's own ancestors and regard the robber as one's father' (Hua Loo-Keng 1951). Hua admitted his lack of historical competence and avoided concern for priority and dating. He chose the most interesting achievements of Chinese mathematics as a simple proof that Chinese people are capable of sophisticated mathematics. His selection included the Pythagorean theorem, the calculation of π, simultaneous congruences, the Pascal triangle and higher-degree equations. The items of this list were obviously selected for their central place in mathematical theory and practice, as well as for the technical sophistication of what was done in these areas. Hua Loo-Keng wanted to show that even though there was now a huge gap between the mathematics of capitalist societies and Chinese mathematics, 'this gap is only temporary and not essential!'

Hua continued to bring up historical achievements of Chinese mathematics in later years. In 1956, the Chinese Mathematical Society organized a mathematical competition for secondary schools on the Soviet model. Hua initiated a series of brochures for the participants of these competitions and other mathematical amateurs, called *A Small Mathematical Series* (*Shuxue xiao congshu* 数学小丛书), which introduced higher branches of elementary mathematics. Four of Hua's contributions to this series were inspired by major achievements of Chinese mathematics of the past. The first one introduced the binomial theorem: *Starting from the Yang Hui triangle* (Hua Loo-Keng 1956). Further volumes talked about approximation methods in numerical mathematics: *Starting from the Zu Chongzhi circle ratio*, i.e. π (Hua Loo-Keng 1962); methods of number theory: *Starting from Liu Hui's section of the circle* (Hua Loo-Keng 1964a); and Diophantine equations *Starting from Sunzi's weird and wonderful calculations* (Hua Loo-Keng 1964b). Hua's search for Chinese achievements was not limited to the ancient period. In 1963, he wrote another popular book about mathematical induction, where he proved by induction the sum formula of the late Qing mathematician Li Shanlan (1810–82) as a complicated example (Hua Loo-Keng 1964c).

Hua also included references to Chinese material in his lectures at the University of Science and Technology of China. The published lecture notes (*Introduction to Higher Mathematics*) mentioned Chinese forerunners when introducing real numbers, limits, the Horner–Ruffini method of numerical solution of higher-degree equations, and even seemingly unrelated topics such as the concept of valid digits. Hua explained how rounding works on the number π and added: 'Zu Chongzhi first showed that $3.1415926 < \pi < 3.1415927$. This is a very good example for valid digits and error, and a glorious contribution to the history of mathematics' (Hua Loo-Keng 1963: 33).

Hua's presentation consciously led to modern standard mathematics and symbolism, and the Chinese material was only an imperfect early stage of these ultimate goals. Historical facts, dating and textual study played no role at all in these simple introductions: their sole purpose was to introduce mathematical topics and raise readers' confidence by stoking their national pride.

Wu also engaged in a similar activity during the Great Leap Forward. His popular introduction to game theory (Wu Wen-Tsun 1959a) included an allusion to Chinese historical literature. Wu illustrated the notion of non-cooperative games by a horse-betting story from the first-century-BCE *Records of the Historian* (*Shi ji* 史记). The clever strategist Sun Bin advised the prince Tian Ji to win a three-round race by matching his best horse against opponent's middle horse, his middle horse against opponent's low-grade horse, and his low-grade horse against opponent's best horse, scoring two wins and one loss.[14]

The demand for national pride in the field of science and technology only became stronger after the Cultural Revolution, and history was immediately called to assist. Hua Loo-Keng's brochures were reprinted in 1978 (clearly also for mathematical-didactic reasons). In 1978, a large volume on all aspects of ancient Chinese science and technology was compiled by the Institute for the History of the Natural Sciences of CAS, called simply *Achievements of Ancient Chinese Science and Technology*. Wu Wen-Tsun's first detailed article (Wu Wen-Tsun 1978a) appeared in this volume. History of Chinese mathematics as a source of confidence also appeared in Xu Chi's celebratory short story (1978) about Chen Jingrun, which concluded:

> We have fallen behind since the Ming dynasty. However, the Chinese apparently have a special talent for mathematics. China should produce great mathematicians. China is the old home of mathematics.
>
> (Xu Chi 1978: 55)

All these examples show a cultivation of a climate in which Wu's historicist investigations were eagerly accepted. Hua Loo-Keng was perhaps a pioneer in this respect.

Wu Wen-Tsun was politically and personally much closer to Guan Zhaozhi than to Hua Loo-Keng, especially after the mid-1960s when Hua left IMCAS for the University of Science and Technology of China. Guan Zhaozhi was primarily interested in the philosophy of mathematics. History was subordinated to this interest, and used to demonstrate the superiority of dialectical materialism. But Guan also strived to keep his articles on a scholarly level, used extensive references and drew on first-hand readings of foreign philosophical articles.

Guan's review of Struik (1948) was his first serious engagement with the history of mathematics. Guan praised Struik's book for its dialectically materialist assumptions, especially for showing links between mathematics and production, class interests and ideology. But he noted critically its biases and inadequate understanding of some sources. According to Guan, Struik should have paid more attention to the 'Chinese mathematicians of the feudal era', but also to nineteenth-century Russian mathematicians, who could problematize Struik's claim that pure and applied mathematics gradually separated in Western Europe in the nineteenth century (Guan Zhaozhi 1956b).

Guan called for others to join the effort to write a better history of world mathematics, but remained largely alone in the field. He published at least 12 historical articles between 1956 and 1965, on topics such as the foundations of geometry in experience (Guan Zhaozhi 1958a), the invention of calculus (Guan Zhaozhi 1958b), and the origins of functional analysis, his own subject (Guan Zhaozhi 1964). Most of his contributions to the ideologically loaded debates about the correct direction of Chinese mathematics were supported by historical data (Guan Zhaozhi 1955, 1957, 1958c, 1959).

The ideological influence of Guan's views is evident in Wu's later writings, but Wu does not seem to have responded directly to any of the actual texts. Wu admitted (interview with the author, 10 July 2010) that he was not very familiar with Guan's work. Wu may not even have read Guan's review of Struik, or at least did not ascribe much weight to it. For example, Guan criticized Struik for referring to all medieval Islamic mathematicians as 'Arabic', but Wu consistently used this label. Guan's influence on Wu's studies was primarily through the initiation of the movement to study ancient Chinese mathematics, and through Guan's well-preserved personal library of historical works.

Guan had also dealt at some length with Engels' characterization of mathematics, discussed at the start of this chapter. Engels described mathematics as the science of 'quantitative relationships and spatial forms'. The standard Chinese translation of 'quantitative' (*shuliang* 数量) raises connotations of numbers (*shu* 数). Wu understood it in this sense when he emphasized that ancient Chinese mathematical style was superior because it combined numbers and shapes. In Guan's (1957) view, however, the dichotomy was misleading because modern mathematics could use more abstract 'quantitative relations' than simple numbers. Engels was only familiar with mathematical objects of low levels of abstraction, i.e. numbers. Guan argued that according to dialectical materialism, the only dichotomy in mathematics should be between quality and broadly defined quantity covering all aspects that can be mathematized, including spatial forms. The objects of modern mathematics are, according to Guan, all types of quantitative relationships in the real world abstracted from concrete qualities. Numbers and shapes are only two important aspects of the subject matter of mathematics, but do not exhaust the range of quantitative relationships (Guan Zhaozhi 1957).[15] In later years, however, Guan did not push his critical reading of Engels any further, and his earlier discussion was ignored or forgotten by all, including Wu Wen-Tsun.

Whereas Hua Loo-Keng contributed to the nationalist climate around Chinese mathematics, Guan provided a model of integration of philosophical concerns with the history of mathematics. Guan's persistent returns to the history of mathematics as a guide for its future direction were shared by Wu from early on.

Construction of traditional Chinese mathematics

Wu's concept of the mechanization of mathematics (discussed in detail in the next chapter) was inspired by a mathematician of Chinese origin, Hao Wang 王浩 (1921–95). Wang, who was based in the USA, visited China in 1972 and 1973–74 and met Wu Wen-Tsun in Beijing. Wang had proved in 1959 all the theorems from Bertrand Russell and Alfred N. Whitehead's *Principia Mathematica* by computer in just over eight minutes, and announced his results with the headline 'Toward mechanization of mathematics' (Wang 1960). In this article, Wang laid out a set of oppositions characterizing calculation on the one hand and theorem-proving on the other: calculations are easy but repetitive, stereotypical and dull, whereas proofs are difficult but short, variable and elegant. This dichotomy merged in Wu Wen-Tsun's thought with the lasting trend in the PRC to denounce mathematical genius and favour practicality and mass accessibility. Turning proof into calculation was a way to remove from the genius the privilege of proving theorems. Wu followed this idea of mechanization back into history and emphasized the mechanizing character of algebraic symbolism, analytical geometry and calculus. These advances had made complicated mathematical arguments accessible to a larger proportion of people, not just to the best mathematical minds as they had been before. Struik's complaint about the Oriental character of high-school mathematics suggested to Wu that Oriental mathematics – and above all Chinese mathematics – could be described precisely as mechanical, as opposed to the Western axiomatic-deductive tradition.

Curiously, Wu Wen-Tsun was not the first writer to call Chinese mathematics 'mechanical', although he was probably not aware of his forerunner. Yoshio Mikami wrote in connection with Song-Yuan dynasty algebra:

> Such a mechanical algebra has no analogies in any other country. Therefore the establishment of Chinese mechanical algebra has a deep significance in the history of mathematics.[16]

> (Mikami 1933: 69)

Mikami applied the label 'mechanization' only to the specific period and domain of Song-Yuan algebra, which involved a lot of repetitive operations with counting rods. Mikami's 'emphasis lies on the evolution of ideas and concepts of a given period and on the features that are considered "specific" to the Chinese tradition' (Horiuchi 2010: xxi). In contrast, Wu turned mechanization into the defining and seemingly permanent feature of traditional Chinese mathematics.

Wu Wen-Tsun's ideas about traditional Chinese mathematics developed further after 1975. He moved beyond a simple listing of its achievements to a summary of basic features under which these achievements belonged. These were most clearly formulated in Wu's entry 'Mathematics' in the *Great Chinese Encyclopedia*:

> Superiority of the calculation methods was beneficial to concrete solution of practical problems. The mathematics that developed from this formed a unique system with constructiveness, emphasis on calculation and algorithms, and mechanization as its features, and problem solving as its chief aim.

> (Wu Wen-Tsun 1988a)

These attributes are all uniformly positive in Wu's philosophy of mathematics. They are in fact modern concepts created for specific modern purposes. This is not to say that they are somehow wrong or necessarily misleading when used to describe traditional Chinese mathematics: rather, the word 'construction' highlights the complete dependence of Wu's generalizations on his perspective and concerns. Wu initially tried to link the history of mathematics and these philosophical concerns with his work on the I^*-functor. But if this was a sincere account of Wu's creative process, he did not later find it convincing, as this claim has never been repeated. The main reason might be that his work on the I^*-functor, although reviewed favourably by his peers, was not seen as a major breakthrough. As soon as Wu found a new passion for algorithmic, or as he called it 'mechanized', mathematics, he could claim Chinese ancestry more plausibly than in algebraic topology. He continued to do so even when such rhetorical shielding was no longer necessary. Let us now investigate the interplay of Wu's mathematics and his historical interests after the end of the CR period.

Notes

1 The identity and affiliation of 'Shu Jin' is unclear; indeed we cannot be sure whether both articles under this pseudonym were written by the same author. The richness of their content and style suggest that they were written by a pre-1966 historian of Chinese mathematics. 'Shu Qun' could be Yuan Xiangdong 袁向东 (born 1942), who later specialized in the history of modern Chinese mathematics. Yuan's early interest was Zu Chongzhi's work on the *Calendar of Great Clarity* (*Da Ming Li* 大明历, interview with the author, 17 October 2008), which is precisely the focus of the Shu Qun article.

2 Wu Wen-Tsun's interview with the author, 10 July 2010. Wu Wen-Tsun's authorship was allegedly immediately clear to those close to him (Li Wenlin 2001), and was revealed in Wu Wen-Tsun (1982c).

3 In later years, Wu expressed his hopes that some such influence could one day be shown (Wu Wen-Tsun 1984c), but sometimes acknowledged the possibility of independent invention, for example of the formula for the quadratic equation by al-Khwarizmi (Wu Wen-Tsun 2003: 22).

4 Used in Gu Jinyong (1975), Wu Wen-Tsun (1976, 1978c, 1979b, 1987b), and also in Wu's entry 'Mathematics' for the *Great Chinese Encyclopedia* (Wu Wen-Tsun 1988a), probably written several years before publication.

5 See, for example, Gray (2007).

6 The 'tables, chairs and beer mugs' quotation originally comes from Blumenthal (1935: 402). For the history of Hilbert's *Foundations of Geometry*, see Toepell (1986).

7 Most readily available scholarly treatments of the Bourbaki phenomenon deal with its philosophy of mathematics and its influence on world mathematical thought: Corry (2004); cf. Aubin (1997). A history focused more on the group and its members has been written as a thesis by Lilian Beaulieu (1989); a planned book by the same author, *Nicolas Bourbaki: History and Legend, 1934–1956*, is apparently not currently available. Her shorter note (Beaulieu 1999) is interesting for its account of the conscious identity-building of the group, which is particularly relevant with respect to Wu.

8 Wu's losses of books were not only due to the Red Guards' harassment but also because of severely limited living space.

9 Struik later received the Chinese translation and was spurred by the critical comments to gain better knowledge of Chinese mathematics. Chinese mathematics was integrated into the main narrative of his book in its third edition in 1969 (Rowe 2001). Wu, however, did not seem to have registered this updated version.

10 It is not clear when this meeting took place. Wu's speech quoted here was given on 22 July 1984, a month *before* the Third International Conference on the History of Chinese Science held in Beijing, where Libbrecht (but not Wu) gave a talk.

11 Wagner (1979) is mentioned in Wu Wen-Tsun (1987b, 1987a). The most prominent reference, however, is in Wu's preface to the *Great Series on the History of Chinese Mathematics*, repeated in each of the eight volumes of the series.

12 Wu mentions Abhyankar's article – but not Abhyankar by name – in Wu Wen-Tsun (1984c). From the same source, we know that Wu had met a student of Abhyankar who was setting up a graduate training centre in China, most probably Tzuong-Tsieng Moh, Abhyankar's colleague from Purdue University.

13 Note that Needham and Wang Ling's list has been criticized point by point in recent Western scholarship (Martzloff 1997: 91).

14 See 'Biography No. 5 of Sun Zi and Wu Qi' in Nienhauser (1994: 39–40).

15 Guan was initially also highly fond of axiomatization and Bourbaki-style mathematical structuralism. Indeed, his abstract view of quantitative relationships directly leads to this position. It is not clear whether he later changed his mind.

16 The first sentence in the Chinese translation: *ru si jixie de daishuxue, ta guo wu qi lei li* 如斯机械的代数学, 他国无其类例. Mikami's book (*Shina no sūgaku no tokushoku* in Japanese) was first published in 1926 and is Mikami's 'best known work (…) fairly representative of his manner of historical writing' (Horiuchi 2010: xxi).

6 Independence and inspiration: Wu Wen-Tsun's work since 1977

Wu Wen-Tsun's modernization of mental labour

One of the last acts of Premier Zhou Enlai 周恩来 (1898–1976) was the reiteration of the policy of Four Modernizations (of agriculture, industry, national defence and science) at the National People's Congress of 1975. The Four Modernizations had been first proclaimed in 1963, but were completely sidelined by the Cultural Revolution. Zhou Enlai and his deputy, Deng Xiaoping 邓小平 (1904–97), raised them again in the mid-1970s as a more pressing issue, with a newly articulated goal to make China a 'modern socialist country' by the end of the twentieth century. The Four Modernizations then became a defining theme of Deng Xiaoping's final victory in the power struggle at the Third Plenum of the Eleventh Central Committee in December 1978.

During these years of Deng's triumph at the end of the 1970s, commitments to help the Four Modernizations were widespread as expressions both of genuine aspiration and of a correct political line. An area which mattered perhaps more than any other to the Four Modernizations was computer science. The development of computers, initiated in China in 1958 but proceeding very slowly during the years of international isolation, suddenly became a hot field drawing the interest of many mathematicians.

Computer research was pursued in several places, including the CAS institutes of Mathematics and of Computation Science, which resided in the same building. IMCAS had a Computation Centre (*jisuan zhan* 计算站), consisting of a Machine Hall (*jifang* 机房) and a Computer Software and Methods Group (*jisuanji ruanjian he fangfa zu* 计算机软件和方法组), and also an affiliated Computer Factory. In 1975, a year for which the records are relatively well preserved in IMCAS archives, the Computations Centre performed over 4,000 hours of computations for organizations from the entire CAS and beyond. The factory produced 40 units of their latest design, Great Wall 203, a large mainframe with 4 kilobytes of memory. And applications of computer methods to mathematical problems were actively promoted throughout IMCAS (IMCAS Party Steering Group 1976).

One of the pioneers of this exploration of the power of computers was Lu Qikeng, Wu Wen-Tsun's long-time colleague, who had in 1972 or 1973 returned to IMCAS from the special IPCAS Research Section No. 13, devoted to criticizing the theory of relativity (see Chapter 4). He was afraid to engage in fundamental research and instead experimented with computer-aided resolution of complex algebraic formulae, and even with theorem-proving (interview with Lu Qikeng, 13 October 2008). Work on computer-aided theorem-proving in IMCAS was mentioned with interest in the report of a US computer science delegation (Cheatham, *et al.* 1973), possibly referring to Lu Qikeng.

Wu Wen-Tsun repeatedly claimed that he came to understand the power of computers earlier during his work at the First Beijing Wireless Telegraphy Factory (*Beijing wuxiandian yi chang* 北京无线电一厂), where he was dispatched to 'engage in labour' in 1970 and 1971 (Hu Zuoxuan and Shi He 2002: 76). In fact Lu Qikeng claims that he introduced Wu to the power of actual computers – he even taught Wu the basics of machine code and practical skills such as registering for machine time. He shared an office with Wu Wen-Tsun and got him interested in the possible use of computers to perform complex symbolic calculations, and finally perhaps even theorems. The Wireless Telegraphy Factory (called the 'Telecommunications Factory in Peking' by Western observers) produced, after all, mostly computer parts, or 'simple, single-purpose production automation devices' (Cheatham, *et al.* 1973). Lu Qikeng's claim that he was crucial for Wu Wen-Tsun's turn to computers is thus quite credible. Ironically, it was Wu who later developed the idea of computer proofs when he, unlike Lu, got CAS funding in 1979 to purchase a much more powerful computer (Hewlett-Packard 9835A with 256 KB of memory) during his visit to USA (Hu Zuoxuan and Shi He 2002: 108–12).

Given the primitive nature of the available machines, Wu first looked for a method that could be tried by hand. He focused on theorems in elementary geometry, drawing on his rich experience with them from his teaching years in wartime Shanghai. Instead of using synthetic Euclidean demonstrations, Wu translated a theorem into a set of polynomial equations, and checked whether the premises and conclusions were algebraically consistent. During the 1977 Chinese New Year holiday, he achieved his first successes (Wu Wen-Tsun 1980). This led to the publication of 'On the decision problem and the mechanization of theorem-proving in elementary geometry' (Wu Wen-Tsun 1977b), Wu's most cited publication on Google Scholar, Web of Science and CNKI. As he realized the need for a more powerful machine for his further experiments, Wu linked his emerging method and research programme to the Four Modernizations, as well as the older concern for independence:

> Mathematical workers of our country should (…) come up with our own viewpoints and methods, establish a couple of mathematical schools with our country's characteristics, produce a series of internationally influential mathematicians, in order to make a sizeable contribution to the solution of mathematical problems posed by socialist construction and to the Four Modernizations independently and from our own initiative.
>
> (Wu Wen-Tsun 1978f: 21)

The emphasis on national independence was linked to a more universal theme. Wu believed that with the development of computers, the laborious, patchy style of axiomatic-deductive reasoning would be superseded by algorithmic, mechanical methods whenever possible:

> Proofs of some theorems or categories of theorems can avoid the common, elegant but mysterious and therefore difficult mode, and use the repetitive but routine and therefore easy mode (…) Computers can thus liberate people from the mental labour of certain logical inferences, and enable mathematicians to use their intelligence on truly innovative work.
>
> (Wu Wen-Tsun 1978f: 21)

China, which had missed the mechanization of physical labour and was 'backward by a huge distance and beaten everywhere' (Wu Wen-Tsun 1984c), should use this opportunity to catch up and assume a position at the forefront of progress:

> Our country has, in the revolutionary mechanization of physical labour, fallen behind several countries that have already become highly developed, and in this present revolutionary mechanization of mental labour, we definitely must not fall into the old track again. We have to try hard to move ahead of all countries of the world.
>
> (Wu Wen-Tsun 1978e: 374)

Wu later elaborated on this theme even further, adding new arguments and details about the utility of mechanization. The ultimate aim of mechanization of mental labour was to liberate mental powers, consumed by pleasant but often useless proving of geometric theorems, to do more demanding tasks for which no solution has yet been found (Wu Wen-Tsun 2002, 2005).

As Wu was arguing for the import of foreign technology, he had to use ingenious arguments to persuade the leadership that it would also, paradoxically, increase independence and self-reliance. The absolute necessity to participate in the 'mechanization of mental labour' in order to prevent another instance of backwardness was exactly such an argument. He foresaw a coming wave of modernization of mental labour, which should be part of the Four Modernizations.

Wu's work was regarded as a marginal curiosity by his Chinese colleagues for a few years, but eventually achieved success and international recognition, and greatly increased his status in China. In the mid-1980s, he took on graduate students and through them established an influential school. But he remained alone among them in linking Chinese tradition to his current research.

From mechanization to equation-solving

The idea of doing mathematics mechanically is often traced to René Descartes and G.W. Leibniz, but it only became workable with the advent of electronic computers in the late 1940s.[1] In 1957, Alfred Tarski first outlined a method to prove theorems of elementary geometry formulated as algebraic equations, and proved that this method would always lead to a decision in a finite number of steps (Tarski 1957). Tarski's method and its later improved versions were, however, of tremendous computational complexity, and no geometrically significant propositions were proved by them. Wu Wen-Tsun greatly simplified the complexity by focusing on theorems without logical quantifiers. These theorems were all logically simple statements about incidence, intersections and distances, which could be expressed as algebraic equations. It was then only a matter of routine algebraic operations to check whether these equations hold universally.

Wu made this vague formulation more precise with concepts of algebraic geometry, where the roots (zeros) of algebraic equations define abstract spaces (algebraic varieties). His first article explained the theoretical feasibility of his method by abstract theorems about the algebraic varieties defined by a theorem in elementary geometry. However, Wu did not present his actual algorithms in this first paper: he only stated the existence of algorithms to do the desired job, showed what such algorithms could look like in the proofs, and added that they would be based purely on elementary algebraic techniques. Although he illustrated his idea with several examples, he solved them using relatively high-level, *ad hoc* techniques (such as substitution) instead of a single low-level algorithm.

Over the following seven years, Wu elaborated the ideas from his first article both theoretically and into fully specified algorithmic instructions. He found a ready-made theoretical framework in the books of Joseph Fels Ritt (1893–1951). Ritt, who spent most of his career teaching at Columbia University, was principally an analyst, interested in 'problems of classical simplicity' and drawing 'inspiration from the great masters of the nineteenth century' (Smith 1956). In 1932, he published the book *Differential Equations from the Algebraic Standpoint* in order to develop what had been a neglected aspect of differential equations, and to 'bring to the theory of systems of differential equations which are algebraic in the unknowns and their derivatives, some of the completeness enjoyed by the theory of systems of algebraic equations' (Ritt 1932: iii).

Because Ritt's theory was created for algebraic differential equations, Wu also suggested an extension of his method to elementary differential geometry in Wu Wen-Tsun (1979a). He also discovered in D. Hilbert's *Foundations of Geometry* (Hilbert 1899), a general method for turning some geometric theorems (pure intersection-point theorems) into algebraic equations. The discovery, made during a lecture course on foundations of geometry in 1978–79, gave him an opportunity to think about the relations between mechanization and the axiomatic systems of different geometries, and to demonstrate his method on easier examples more transparently (Wu Wen-Tsun 1982b). A full theoretical and practical exposition of the method, now called Wu–Ritt or CharSet because of its reliance on Ritt's concept of characteristic sets, was made in the article 'Basic principles of mechanical theorem proving in elementary geometries' (Wu Wen-Tsun 1984a) and in the Chinese book *Basic Principles of Mechanical Theorem Proving in Geometries* (Wu Wen-Tsun 1984b). By this time, Wu had also written a computer program to test his method and run it on the HP computer that he had bought in the USA. Several non-trivial theorems could be proved with surprising speed, e.g. Simson's line theorem, Feuerbach's theorem about the nine-point circle, and even theorems in non-Euclidean elementary geometries (e.g. hyperbolic geometry).

After 1984, Wu's work, assisted by graduate students, proceeded along two lines. He continued to write on the theoretical foundations of his method belonging to algebraic geometry (Wu Wen-Tsun 1986a, 1989b). On the other hand, he applied his improved method to theorems in elementary geometry (Wu Wen-Tsun 1987d, 1987e), to the derivation of Newton's from Kepler's laws (Wu Wen-Tsun 1987f), and to chemical equilibrium problems (Wu Wen-Tsun 1990a). Later, he and his students managed to expand the method, via differentiation, to the solution of inequalities (Wu Wen-Tsun 1988b) and optimization problems (Wu Wen-Tsun 1992).

Wu's work on mechanization of mathematics is highly cited in all databases surveyed, although not every publication received the same amount of attention. Even granting that citation counts from different periods and fields cannot be directly compared, it seems safe to say that Wu's post-1977 research has attracted more readers than his earlier work on pure mathematics (Table 6.1).

References to Wu often also cite Chou Shang-Ching (1988), which presented Wu's theory in a comprehensive and accessible way. Chou Shang-Ching (S.C. Chou, Zhou Xianqing 周咸青, born 1942) entered CAS Graduate College in 1978 and attended Wu's lectures, but did not approach the famous mathematician personally at the time. In 1981, he enrolled at the University of Texas, Austin, and joined the computer science group of R. Boyer and J. Moore there. Chou introduced what he remembered about Wu's theory at Boyer's seminar, which incited the interest of another UT computer scientist W.W. Bledsoe. Bledsoe asked Chou and his colleague Wang Tie-Cheng 王铁城 to procure relevant articles from Wu Wen-Tsun, and Wu quickly sent them the only two papers he had in English at this point, his first article (Wu Wen-Tsun 1978d) and a conference paper (Wu Wen-Tsun 1981b). Bledsoe, Boyer and Moore felt that the actual computation was not clear from the papers, and asked Chou to reconstruct it and turn it into a true computer algorithm. Parts of Chou's 'reconstruction', especially the triangulation of polynomials (see below), had to be created with hardly any clue from Wu's articles. Chou nevertheless succeeded surprisingly quickly, and reported his success back to Wu in September 1982. Chou announced his results at the American Mathematical Society Annual Meeting in Denver in January 1983, and received many requests for copies of Wu's original articles. Hundreds were sent out all across the USA in 1983–84. The original article 'On the decision problem …' was reprinted in Bledsoe and Loveland (1984), accompanied by

Table 6.1 Wu Wen-Tsun's five most-cited post-1977 publications, in chronological order

Article	MathSciNet	CNKI	WoS/Scopus	Google Scholar
Wu Wen-Tsun (1977b) (Chinese) / Wu Wen-Tsun (1978d) (English)	21	43	70	336
Wu Wen-Tsun (1984b)	22	6	55*	436
Wu Wen-Tsun (1985b) (Chinese) / Wu Wen-Tsun (1986a) (English)	16	25	70	129
Wu Wen-Tsun (1989b)	10	N/A	N/A	107
Wu Wen-Tsun, *et al.* (1994) (book)	14	N/A	N/A	161

*Citations in Scopus on a 1986 reprint of the same paper in the *Journal of Automated Reasoning*. All other citation counts in this column are from the Web of Science.

Chou's more accessible Denver talk. It was Chou who coined the term 'Wu's method' (Chou Shang-Ching 2010).

The initial acceptance of Wu's method abroad relied intimately on networks of scholars of Chinese descent, as well as on the growing interest in computer science in the USA (much stronger than in China, where few people had any experience with computers). These networks were also essential for arranging Wu's visits to America in the 1980s.

Wu's success in theorem-proving inspired, among others, a mathematician in China, Zhang Jingzhong 张景中 (born 1936). Zhang was expelled from Peking University during the Anti-Rightist Campaign and worked manually in the countryside from 1957 till 1978, but he returned to mathematical research whenever he could. In 1978, he joined the Department of Mathematics at USTC in Hefei, and noticed Wu's first article on mechanization. When he later joined a new Institute of Mathematics of the Chengdu Branch of the Chinese Academy of Sciences, he designed his own method of computer proofs, using a different principle than Wu (experimental verification of the theorem by a suitable number of 'random' points), but in later years he collaborated with the Wu group (Zhang Jingzhong 2010).

The changing weight of Chinese and non-Chinese recipients of Wu's publications can be illustrated by a breakdown of the citations of his first article, as retrieved from Google Scholar. The 'Chinese' category includes Chou Shang-Ching, who published mostly abroad in the early years and only towards the end of the 1980s started turning out collaborative articles with Wu Wen-Tsun's students. If we changed the affiliation of these early works by Chou to 'non-Chinese', it would make clearer the prominent role of foreign recognition of Wu's work for his acceptance in China (Figure 6.1).

In 1979, Wu transferred from IMCAS to the newly established Institute of Systems Science (*Xitong kexue yanjiusuo* 系统科学研究所, ISSCAS).

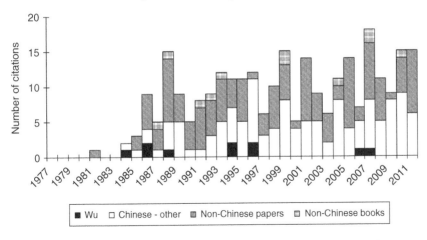

Figure 6.1 Structure of citations of Wu Wen-Tsun (1978d) in *Google Scholar*

The creation of this institute was to a large extent a result of the Cultural Revolution, which added a dimension of personal hatred to relationships previously strained by scholarly disagreement. Guan Zhaozhi first suggested the creation of a new institute devoted to modern applied mathematics to avoid further conflicts with Hua Loo-Keng. But because Hua had worked since 1964 exclusively on applied mathematics, IMCAS ended up divided into three parts. Alongside the residual Institute of Mathematics and a new Institute of Applied Mathematics, both formally headed by Hua Loo-Keng and dominated by his students, the Institute of Systems Science was formed on 29 July 1979, headed by Guan Zhaozhi (Wang Yuan 1999: 370–73). The official CAS report on the foundation of the new institutes defined their tasks as follows:

> The Institute of Applied Mathematics will mainly pursue studies into mathematical methods and fundamental theories serving the national economy, national defense and other scientific disciplines. The Institute of Systems Science will mainly pursue studies of control theory and overall [*zonghe* 综合] theory of large systems in the national economy and national defense, as well as fundamental research in related border disciplines.
>
> (Chinese Academy of Sciences 1979)

Due to the largely personal rationale for the division, there was a significant overlap in both pure and applied disciplines studied in all three institutes.[2] Wu Wen-Tsun was thus able to pursue his new interest, mechanization of mathematics, in ISSCAS, even though it did not really fall within the scope defined above. He received Guan's unreserved support and was also backed by the ISSCAS Party Committee, which accepted him into the CPC in 1980 (Hu Fanfu 1980).

During the 1980s, Wu travelled to West Germany and France as well as the USA, and published his last articles and a monograph on the *I**-functor. But he did not supervise any further students in topology, and was clearly determined not to return to it (Li Banghe 2010).

During the 1980s, his interest gradually moved from narrowly defined theorem-proving to broader 'mechanization of mathematics', including general applications of computer-based algebraic methods to deal with various geometric problems. In 1990 he established a Centre for Mechanization of Mathematics in ISSCAS. His research agenda and results became more recognized and institutionalized. His students, especially Gao Xiaoshan 高小山 and Wang Dongming 王东明, formed a productive research school within the project 'Machine proofs and their applications', designated key research projects in 1992 and generously funded by the government. It was expanded into a 'Forum for Mathematics Mechanization and Automated Reasoning' in 1998, headed by Gao Xiaoshan (Hu Zuoxuan and Shi He 2002: 158–62).

The period 1985–90 witnessed a surge of enthusiasm, in China and abroad, for computer proofs based on Wu's method. Although this dispelled the earlier doubts about its viability, it also brought a proliferation of alternative approaches based on fancy new mathematics from abroad. Wu's students were keeping track of worldwide developments in their field and had no time for ancient Chinese mathematics (Wang Dongming, *et al.* 2010).

Gradually, the stated goal of Wu's circle moved from a grand vision of mechanized proofs to a more modest and mainstream programme of solving large systems of equations by computer, and applying this technique to diverse fields in pure mathematics, artificial intelligence (computer-aided design and drawing, pattern recognition, etc.). Wu had turned into the founder of a vigorous research agenda, which by the second decade of the twenty-first century had completely freed itself of his initial interests and historical-philosophical motivations. What remained, however, is the interest in computation, arguably the most durable and visible feature of Chinese mathematics in both the traditional and modern period.

Wu's method – brief description

In the rest of this chapter, I will try to make more precise the often-repeated claim that Wu drew on traditional Chinese mathematics. In order to do that, it is necessary to explain the structure of Wu's method, the algebraic techniques used within it, and the role they play there.

Wu's method of geometric theorem-proving proceeds in four steps:

1 Algebraization – Formulation of algebraic equations corresponding to the geometric proposition.
2 Triangulation – Construction of an equivalent set of equations of gradually increasing numbers of variables.

3   Elimination – Pseudo-division of the conclusion polynomial by the trian-
     gulated premises to eliminate all dependent variables.
4   Geometrization – Interpretation of the result and of auxiliary conditions
     generated by step 3).

In the first step, the prover constructs algebraic equations expressing the premises and conclusion of a geometric theorem. The variables of these equations
are usually coordinates of important points, which are linked by the statement
of the theorem. This follows from Wu's limitation of his method to elementary
geometry with statements 'which can be formulated and established without
the help of any set-theoretical devices', i.e. statements about finite numbers of
points (Tarski 1959). Geometric properties (distances, incidences of points and
lines, etc.) are 'translated' into algebraic relations between the chosen variables.
Statements about points then correspond to equations about their coordinates.
In most theorems of elementary geometry, these equations will be polynomial
or easily transformable to polynomial equations of the form $P = 0$. Wu gave
the following example of the most common algebraizations:
    For example, for points $A_i = (x_i, y_i)$, distinct or not, we shall say:

> A_1A_2 is parallel to A_3A_4
> if $(x_1-x_2)(y_3-y_4) - (x_3-x_4)(y_1-y_2) = 0$,
> A_1A_2 is orthogonal to A_3A_4
> if $(x_1-x_2)(x_3-x_4) + (y_1-y_2)(y_3-y_4) = 0$,
> the length-square of A_1A_2 is $r^2 = (x_1-x_2)^2 + (y_1-y_2)^2$, etc.
> (Wu Wen-Tsun 1978d: 215)

The coordinates have to be arranged in a sequence and classified into two
groups. As there are always fewer premises (m) than coordinates (n), the first
$n - m$ coordinates have to be treated as independent parameters (denoted by
u_i). The last m coordinates (denoted by x_i) are treated as algebraically fixed by
the m premises and dependent on the parameters. The choice of variables, their
ordering, and division into parameters and dependents, are non-trivial tasks
that require some preliminary understanding of the geometrical proposition,
otherwise the results might be geometrically meaningless. Algebraization was
thus only partly mechanized in the first version of Wu's method.
    If the proposition to be proved is valid, the equation corresponding to the
conclusion must be true for any values of the independent parameters. This
can be checked by eliminating all dependent variables from the conclusion
polynomial. If the resulting polynomial in parameters only is a tautology, the
proposition holds. In fact, elimination will impose some additional algebraic
constraints on the parameters, called non-degeneracy conditions. In Wu's formulation, the task is:

> Given a system Σ of polynomial equations (or equivalently, system of
> polynomials) as well as another polynomial g, all in the same finite set

of variables x, y, ..., decide in a finite number of steps either of the two cases below:

Case 1. A finite set of polynomials D_a is determined such that $g = 0$ is a consequence of the system Σ under the non-degeneracy conditions $D_a \neq 0$ such that D_a are themselves not consequences of the system Σ.

Case 2. No such set $S = \{D_a\}$ can exist so that Case 1 holds.

(Wu Wen-Tsun 1984a: 208)

The conclusion g will be a consequence of Σ when it can be expressed as algebraically dependent on the system of premises Σ and the system of non-degeneracy conditions D:

$$D_1 D_2 \ldots D_n g = Q_1 \Sigma_1 + Q_2 \Sigma_2 + \cdots + Q_n \Sigma_n \qquad (6.1)$$

Equation (6.1) holds if g can be successfully reduced by elimination. Elimination is the core of Wu's method, and the locus of the link to traditional Chinese mathematics.

Elimination is analogous to the Euclidean division algorithm familiar for integers and polynomials. For any two integers $n \geq m > 1$, there is precisely one integral quotient q ($q < n$) and one integral remainder r ($r < m$) such that:

$$n = qm + r \qquad (6.2)$$

For example, if $n = 8$ and $m = 3$, the division would be $8 = 2 \cdot 3 + 2$. Similarly, for two polynomials $N(x)$ and $M(x)$ of a real variable x, with degrees (the highest power of x) $\deg(N) \geq \deg(M) > 0$, the quotient and remainder polynomials can be found:

$$N(x) = Q(x)M(x) + R(x) \qquad (6.3)$$

In an analogy to the integral situation, $\deg(Q) < \deg(N)$, $\deg(R) < \deg(M)$. Again an example: if $N = x^2 + 2x + 3$ and $M = x + 3$, the division gives $N = (x - 1)M + 6$.

In both the integral and single-variable polynomial case, the quotients and remainders are also either integers or single-variable polynomials. This means that it is possible to divide the remainder using the same algorithm by another, smaller, m or M until eventually we reach a remainder that is either zero (the original n or N is fully divisible), or smaller than any divisor (n or N are not divisible). This is not the case with polynomials in several variables $F(x_1, \ldots, x_n)$ and $G(x_1, \ldots, x_n)$. Their quotients and remainders are often rational functions rather than polynomials, and there is therefore rarely any strict divisibility. A modification of the division algorithm, pseudo-division, solves this problem. To avoid rational functions, both sides of the equation have to be multiplied by the denominator $f(x_1, \ldots, x_p)$:

$$f(x_1, \ldots, x_p) G(x_1, \ldots, x_n) = Q(x_1, \ldots, x_n) F(x_1, \ldots, x_n) + R(x_1, \ldots, x_n) \quad (6.4)$$

The denominator must, of course, be a function of fewer variables than the (pseudo-)dividend and divisor. Again we assume that in x_n, $\deg(G) \geq \deg(F) > 0$, and then it must be the case that correspondingly $\deg(Q) < \deg(G)$, $\deg(R) < \deg(F)$. If $\deg(F) = 1$, $\deg(R)$ must be 0, and the variable x_n has thus been eliminated. If $\deg(R) > 0$ and there is no divisor which could be used to lower this degree, the conclusion is clearly not even pseudo-divisible by the given divisors, i.e. does not follow from the premises.

In this way we can construct a pseudo-remainder of the conclusion g with respect to all polynomials Σ equivalent to the premises. For the implication from the premises to the conclusion to hold, this final pseudo-remainder must be identically zero, making the conclusion an algebraic combination of the premises, thus zero whenever the premises are all zero, in other words for all values of variables permitted by the premises.

Wu's method – an example

Let us demonstrate the technique on a simple example from Chou Shang-Ching (1988). Chou's notation will be adopted throughout, except for a few additional distinctions acknowledged when made.

A trivial theorem in plane geometry states that the diagonals of a parallelogram $ABCD$ always intersect at their midpoints (Figure 6.2). The premises of the theorem are that the opposite sides of a parallelogram are parallel, and that its diagonals have an intersection. The conclusion is that this intersection is the midpoint of both diagonals, in other words the distances from both ends of the diagonal to the intersection are equal. There are actually two statements of this conclusion, one for each diagonal.

These conditions correspond to the following equations:

$$h_1 \equiv (x_B - x_A)(y_C - y_D) - (x_C - x_D)(y_B - y_A) = 0$$
$$h_2 \equiv (x_C - x_B)(y_D - y_A) - (x_D - x_A)(y_C - y_B) = 0$$

$$h_3 \equiv (x_C - x_A)(y_E - y_A) - (x_E - x_A)(y_C - y_A) = 0$$
$$h_4 \equiv (x_D - x_B)(y_E - y_B) - (x_E - x_B)(y_D - y_B) = 0$$

$$g_1 \equiv (x_C - x_E)^2 + (y_C - y_E)^2 - (x_E - x_A)^2 - (y_E - y_A)^2 = 0 \quad (6.5)$$
$$g_2 \equiv (x_D - x_E)^2 + (y_D - y_E)^2 - (x_E - x_B)^2 - (y_E - y_B)^2 = 0$$

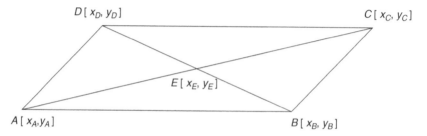

Figure 6.2 Points of a parallelogram and their coordinates

Note: The labels here differ from Chou Shang-Ching (1988).
H_1: *AB* and *CD* are parallel.
H_2: *BC* and *AD* are parallel.
H_3: *E* lies on *AC*.
H_4: *E* lies on *BD*.
C_1: *AE* is congruent (of equal length) to *EC*.
C_2: *BE* is congruent to *ED*.

Now we choose free parameters (u_i) and dependent variables (x_i). This is an important decision which has to be made in accordance with the inner logic of the theorem. Put simply, independent coordinates should be introduced when a point is arbitrary, dependent coordinates when it is constrained (geometrically constructed). In most constructions, some coordinates of the point will be arbitrary and others constrained. Some parameters can be eliminated at the outset by the choice of the frame of reference. In our example, A will be the origin, B a point on the *x*-axis, C a fully arbitrary point and D and E fully constrained points, with coordinates A [0, 0]; B [u_1, 0]; C [u_2, u_3]; D [x_2, x_1]; E [x_4, x_3]. The system of equations (6.5) then becomes neater:

$$h_1 \equiv u_1 \left(u_3 - x_1 \right) = 0$$
$$h_2 \equiv x_2 u_3 - x_1 \left(u_2 - u_1 \right) = 0$$
$$h_3 \equiv \left(x_4 - u_1 \right) x_1 - x_3 \left(x_2 - u_1 \right) = 0 \qquad (6.6)$$
$$h_4 \equiv x_4 \left(u_3 - x_3 \right) - x_3 \left(u_2 - x_4 \right) = 0$$
$$g \equiv x_3^2 + x_4^2 - \left(u_2 - x_4 \right)^2 - \left(u_3 - x_3 \right)^2 = 0$$

Note that variables (x's) are indexed in the order of their appearance in the premises. It is obvious that $x_1 = u_3$, but we will keep the distinction to highlight the mechanical procedure.

It is now necessary to reorganize the hypotheses so that they have an ascending number of dependent variables. The resulting system will be roughly triangular, hence this step is called triangulation. We see that the system (6.6) already is almost triangular, only h_3 needs to be replaced with a polynomial free of x_4. The pseudo-remainder of h_3 with respect to h_4 is exactly the polynomial needed. We can thus introduce the procedure of elimination on the pair of polynomials h_3 and h_4, aiming to eliminate x_4.

Recalling (6.3), we want to find a similar equation:

$$d(u_1,\ldots,x_3)h_3(u_1,\ldots,x_4) = q(u_1,\ldots,x_3)h_4(u_1,\ldots,x_4) + r(u_1,\ldots,x_3) \quad (6.7)$$

The polynomials d and q are the coefficients of x_4 in h_3 and h_4, respectively. This means that the pseudo-remainder r, written $prem(h_3, h_4, x_4)$, is defined by this equation:

$$u_3 h_3 = x_4 h_4 + prem(h_3, h_4, x_4) \quad (6.8)$$

and calculated as follows:

$$
\begin{aligned}
prem(h_3, h_4, x_4) &= u_3 h_3 - x_1 h_4 \\
&= u_3 x_1 x_4 - u_3(x_2 - u_1)x_3 - u_1 u_3 x_1 - (u_3 x_1 x_4 - u_2 x_1 x_3) \\
&= (u_2 x_1 - u_3 x_2 + u_1 u_3)x_3 - u_1 u_3 x_1
\end{aligned}
\quad (6.9)
$$

As we replace h_3 by $prem(h_3, h_4, x_4)$, we re-label the polynomials f_1 to f_4 and turn the conclusion into the form of a polynomial with ordered variables:

$$
\begin{aligned}
f_1 &\equiv u_1 x_1 - u_1 u_3 = 0 \\
f_2 &\equiv u_3 x_2 - (u_2 - u_1)x_1 = 0 \\
f_3 &\equiv (u_3 x_2 - u_2 x_1 - u_1 u_3)x_3 + u_1 u_3 x_1 = 0 \\
f_4 &\equiv u_3 x_4 - u_2 x_3 = 0 \\
g &\equiv 2u_2 x_4 + 2u_3 x_3 - u_3^2 - u_2^2 = 0
\end{aligned}
\quad (6.10)
$$

This set can be used to eliminate, one by one, the x's from the conclusion g. The technique of pseudo-division is always the same: first cross-multiply the pseudo-dividend and pseudo-divisor by their leading coefficients, then subtract:

$$
\begin{aligned}
prem(g, f_4, x_4) &= u_3(2u_2 x_4 + 2u_3 x_3 - u_3^2 - u_2^2) - 2u_2(u_3 x_4 - u_2 x_3) \\
&= (2u_2^2 + 2u_3^2)x_3 - u_3^3 - u_2^2 u_3 = r_3
\end{aligned}
\quad (6.11)
$$

$$
\begin{aligned}
&prem(r_3, f_3, x_3) \\
&= (u_3 x_2 - u_2 x_1 - u_1 u_3)\left[(2u_2^2 + 2u_3^2)x_3 - u_3^3 - u_2^2 u_3\right] \\
&\quad -(2u_2^2 + 2u_3^2)\left[(u_3 x_2 - u_2 x_1 - u_1 u_3)x_3 + u_1 u_3 x_1\right] \\
&= (-u_3^4 - u_2^2 u_3^2)x_2 + u_3(u_2 u_3^2 - 2u_1 u_3^2 + u_2^3 - 2u_1 u_2^2)x_1 + u_3^4 u_1(u_2^2 + 1) = r_2
\end{aligned}
\quad (6.12)
$$

Skipping further intermediary stages, the result is

$$prem(r_1, f_1, x_1) = u_1^2 u_2^2 u_3^2 + u_1^2 u_3^5 - u_1^2 u_2^2 u_3^3 - u_1^2 u_3^5 = 0 \tag{6.13}$$

The theorem is thus proved. Although this example is very simple, it already leads to some large polynomials. Wu's method generates polynomials with hundreds of terms even for theorems with about ten premises. It is, however, in many cases even today the most computationally efficient algorithm for elimination of variables from polynomial equations.[3]

Wu's elimination and traditional Chinese mathematics

Wu Wen-Tsun repeated on several occasions that his method was inspired by traditional Chinese mathematics. Even in his first article on mechanical theorem-proving, he mentioned the connection between his method and algebraic techniques used in China in earlier times:

> The algorithm we use for mechanical proofs of theorems in elementary geometry mainly involves some applied techniques for polynomials, such as arithmetic operations and simple eliminations of unknowns. It should be pointed out that these were all created by Chinese mathematicians in the 12–14 century Song and Yuan periods, and were already well developed then. The work of Qian Baocong can be consulted for detailed explanation.
>
> (Wu Wen-Tsun 1977a: 516)

He went further in an article for popular audiences in 1980, when he emphasized that although David Hilbert (1899) and Alfred Tarski (1957) had already made some progress towards a mechanical method, he was initially unaware of their results: 'We set out the question and came up with a method of solution under inspiration from Chinese ancient algebra' (Wu Wen-Tsun 1980: 364). But he did not pinpoint any single work or method that had inspired him. Instead, he summarized the history and achievements of Chinese algebra, mentioning in a single sentence mathematicians of the Song and Yuan periods (tenth to fourteenth centuries):

> Concepts corresponding to modern polynomials were introduced, calculation rules for polynomials and algebraic tools for elimination were established, and this brought systematic development to the method of algebraization of geometry, as can be seen in many works of Yang Hui, Li Ye, and Zhu Shijie, fortunately preserved until today.
>
> (Wu Wen-Tsun 1980: 365)

Later, Wu indicated that he engaged in more detail with Zhu Shijie's algorithm for systems of polynomial equations:

The author has learned that some articles were published last year, which improved the currently most efficient methods [for systems of polynomial equations] and tested them on a computer. These tests included at least two seemingly simple systems of equations, one of which could not be solved at all, while the other required huge memory capacity and machine time. This author used a method of his own to easily solve these systems of equations even on a much smaller machine. In fact this 'method of his own' is nothing else than a modernized, generalized form of Zhu Shijie's method of Four Unknowns.

(Wu Wen-Tsun 1987b: 39)

The method of Four Unknowns (literally 'four elements', referring to Heaven, Earth, Man and Matter, *tian di ren wu* 天地人物) is recorded in the *Jade Mirror of Four Unknowns* (*Si yuan yu jian* 四元玉鑒), published by Zhu Shijie 朱世杰 in 1303.[4] On this occasion, Wu did not yet openly declare that this 'method of his own' is the technique used in his method of mechanization. But Li Wenlin quotes him very much to that effect in 2001:

My method of solving equations is basically derived from Zhu Shijie. He used elimination to remove variables one by one, which provided me with a basic model for the method. Of course, Zhu Shijie had no theory, it was very crude, only calculations. I developed it and put it on a truly modern mathematical ground, i.e. algebraic geometry.

(Li Wenlin 2001: 59)

In order to test Wu Wen-Tsun's claim, we will now compare Zhu Shijie's method to Wu's technique of pseudo-division. We will then also evaluate additional evidence of his knowledge of Zhu's techniques at the time when he was designing his method of mechanization.

What should be mentioned first is that the method of Four Unknowns was developed by gradual expansion of the method of two unknowns, introduced by a certain Li Dezai 李德載, according to the prefaces to Zhu Shijie's book. The method of two unknowns was based on a clever notational device intimately connected to the calculation routines of traditional Chinese mathematics. Equations in a single unknown (the 'heavenly element' *tian yuan* 天元) were laid out as a column of coefficients from the constant to the highest power. The method of two unknowns used the two-dimensionality of the calculating surface to express equations of both 'heaven' and 'earth' (*di yuan* 地元 – see Table 6.2).

This notation made it very simple to multiply by the unknowns: the corresponding coefficient was simply shifted down or to the left. The method of two unknowns used this convenience to derive, in a truly mechanical fashion, a single column (equation in a single unknown) from two equations in two unknowns, called 'the left form' (*zuo shi* 左式) and 'the right form' (*you shi* 右式). It was this technique which foreshadowed the crucial elimination step of Wu's method.

Table 6.2 Layout of an equation of a single unknown *tian* and of two unknowns *tian* and *di*

Const.	...	di^2	di	Const.
tian	...	tian*di^2	tian*di	tian
$tian^2$		...	$tian^2$*di	$tian^2$
$tian^3$			...	$tian^3$
...				...

The later expansion of the method of two unknowns by Zhu Shijie's prede-cessors and Zhu himself enabled the solution of systems of equations of three or four unknowns, but in those more complex equations, it was no longer possible to multiply simply by shifting on the calculating surface. The two-dimensional calculating surface did not provide enough space for mixed coef-ficients in three or four dimensions, which had to be eliminated by ingenious substitutions and other *ad hoc* tricks. Zhu Shijie's innovations were thus iron-ically a major step away from mechanical tendencies in Chinese mathematics; but since Zhu Shijie's book was the only place which recorded the method of two unknowns, integrated into his own method of four unknowns, Wu spoke of 'Zhu Shijie's' method. 'Zhu Shijie' and 'Four Unknowns' should, however, be understood as imprecise labels.

All that is actually recorded in Zhu Shijie's original text are numerical con-figurations at the initial, intermediate and final stages of the procedure, punc-tuated by terse slogans, such as *hu yin tong fen* 互隐通分 (the hidden ones are brought on the same denominator), *liang wei xiang xiao* 两位相消 (the two positions cancel each other out) and *nei wai hang xiang cheng* 内外行相乘 (outer and inner columns multiply each other).

Let us illustrate it on the third exemplary problem in the introductory chapter (*juan shou* 卷首) of the *Jade Mirror*. This problem involves three unknowns, but we will start once the third unknown has been eliminated. The resulting configurations are called 'the Former form' (*qian shi* 前式) and 'the Latter form' (*hou shi* 后式), corresponding to $y^2(1-x) + y(x^2+x+1) - 2x^2 - x - 2$ and $y^3 + y^2(-2x-2) + y(x^2+4x+2) - 2x^2 - 2x$, respectively:

Former form			Latter form			
1	1	−2	1	−2	2	0
−1	1	−1		−2	4	−2
	1	−2			1	−2

Now Zhu Shijie 'brings to the common denominator the mutually hid-den [terms]' (*hu yin tong fen*), and then 'subtracts them' (*xiang xiao*). The vocabulary mirrors the procedure of subtracting fractions. Zhu only records the result of these calculations, two forms ('Left' and 'Right') of a lower degree in *y*:

Left form		Right form	
7	−6	13	−14
3	−7	11	−13
−1	−3	5	−15
	1	−2	−5
			2

In symbolic terms,

$$Left \equiv y\left(-x^2 + 3x + 7\right) + x^3 - 3x^2 - 7x - 6$$
$$Right \equiv y\left(-2x^3 + 5x^2 + 11x + 13\right) + 2x^4 - 5x^3 - 15x^2 - 13x - 14 \tag{6.14}$$

How were these results derived? This is not indicated in Zhu Shijie's text, but later commentaries suggest a sequence of steps amounting to the following formulae:

$$Left = (1-x)\left[(1-x)\,Latter - y \cdot Former\right] - \left(x^2 - x - 3\right)Former$$
$$Right = (1-x)\,Left - \left(-x^2 + 3x + 7\right)Former \tag{6.15}$$

The repeated factor (1−x) is clearly the leftmost column of *Former*. The factor (x^2-x-3) turns out to be the leftmost column of the configuration (1−x)*Latter* − y·*Former*. Similarly, the factor $(-x^2+3x+7)$ is the leftmost column of *Left*. In other words, the forms are mutually multiplied by their leftmost columns, the coefficients of the highest power of *y*. Let us call (1−x) the *Initial* (of *Former*); then the following is true:

$$Initial^2\,Latter = \left(y - xy + x^2 - x - 3\right)Former + Left$$
$$Initial \cdot Left = \left(-x^2 + 3x + 7\right)Former + Right \tag{6.16}$$

This means that *Left* and *Right* are successive pseudo-remainders of the pseudo-division of *Latter* by *Former*, as in equation (6.3) of Wu's method. Moreover, the iterative nature of the algorithm, where new remainders replace the original dividends, is precisely the process by which Wu reduces the power of polynomials in his method.

This structural affinity of the method of two unknowns to Wu's method is thus beyond doubt. But was Wu Wen-Tsun familiar with the ancient method at the time when he created his method of mechanization?

The earliest evidence of Wu Wen-Tsun's direct reading of Zhu Shijie's works appeared in a semi-popular article (Wu Wen Tsun 1982c). Wu referred to problem 4 of the chapter 'Right angles and surveying' (*Gou gu ce wang*

勾股测望）, and included the detailed working (*cao* 草) from Luo Shilin's 罗士琳 edition of Zhu Shijie's work of 1834 (*The Detailed Workings of the Jade Mirror of Four Unknowns, Si yuan yu jian xi cao* 四元玉鑑细草）. This is significant because Luo worked out all the intermediate calculation steps of Zhu's method for all the problems in the book. Although this particular problem does not involve a system of polynomials, Wu most likely also read other parts of Luo's commentary.

But even before reading *The Jade Mirror* in the original, Wu was aware of the main idea of Zhu's method from the clear exposition in the history of Chinese mathematics by Qian Baocong (1964). This book was among the references of his articles on mechanization (Wu Wen-Tsun 1977a) as well as on the history of Chinese mathematics (Gu Jinyong 1975). Qian's book includes a particularly lucid description of the elimination process of the method of Four Unknowns:

> 'Bringing to the common denominator the mutually hidden terms and subtracting them' is a procedure of elimination for general systems of equations in two unknowns (…) For example tables with three columns (or polynomials with a square of y) would in general be denoted like this:
>
> $$A_2 y^2 + A_1 y + A_0 = 0 ....(1)$$
> $$B_2 y^2 + B_1 y + B_0 = 0 ....(2)$$
>
> The upper-case letters represent polynomials without y.
>
> If we want to eliminate the term with y^2, we can multiply all terms in (2) apart from $B_2 y^2$ by A_2, then multiply all terms in (1) apart from $A_2 y^2$ by B_2, subtract them from each other, and we get
>
> $$C_1 y + C_0 = 0 ....(3)$$
>
> Then (3) is multiplied by y and set again against (1) or (2), and by the same method of elimination we get
>
> $$D_1 y + D_0 = 0 ....(4)$$
>
> (3) and (4) are then 'two-column forms with two unknowns', from which all members with y can be eliminated by the preceding method.
> (Qian Baocong 1964: 183–4, abridged)

Qian used the same problem that we have discussed (problem 3 of the introductory chapter) to illustrate the method. The same equation system was later used by Wu in a conference paper about old Chinese methods of equation-solving (Wu Wen-Tsun 1993). It was thus probably through the mediating influence of Qian Baocong that Wu realized the potential for elimination of polynomials by this old Chinese method, and when he approached the text of the *Jade Mirror* directly, he did it with an understanding due to Qian. The knowledge that such a method had been used with success in China was in turn a boost to his interest in mechanical theorem-proving in two respects: he had a clue about how to proceed, albeit there was much more work to be done to make it viable both theoretically and as a computer algorithm; and he felt morally strengthened by having a *Chinese* inspiration in particular.

But Wu also explained his method theoretically, using J.F. Ritt's concepts. Since Ritt's work was by the mid-1970s also part of the history of mathematics, it is instructive to look at Wu's use of this theory. One gets the impression from this encounter that Wu was attracted by the simplicity of early twentieth-century mathematics and its simpler concepts and theories.

J.F. Ritt's theory and constructive mathematics

J.F. Ritt's theory of basic (or characteristic) sets is a comprehensive treatment of the reducibility of algebraic varieties (originally differential algebraic varieties). This is a crucial theoretical problem for Wu's method, since it can only work reliably on systems of equations which define irreducible varieties (varieties which are not composed other varieties).

J.F. Ritt's theory was introduced in an article (Ritt 1930) and then a book (Ritt 1932). He analyzed differential forms, formal equations similar to polynomials, but with the added dimension of orders of derivatives. Coefficients and variables in these forms are differentiable functions:

$$\left(\frac{d^q y_p}{dx^q} \right)^r$$

Ritt's ultimate aim was to decide which systems of differential forms are irreducible and reduce those that are reducible. He found an iterative method of constructing irreducible systems from what he called 'basic sets' of differential forms.

We will explain the notion of basic sets, as adopted in Wu's method, for the purely algebraic (non-differential) case, where each polynomial is composed of elementary forms $x_p{}^q$. Here, p, defined by some ordering of variables, is the *class*, and q the *degree* of the form. The class of a polynomial is the highest class of its forms.

A set of polynomials A_i is called an *ascending set* A if each subsequent polynomial has a class higher than the preceding polynomial (introduces a new variable), and if the newly introduced variable only appears in lower degrees in all subsequent polynomials. Ascending sets can be partially ordered by a comparison of ranks (classes or degrees) of their corresponding polynomials. A finite system of polynomials Σ defines various ascending sets; those with the lowest ranks are called *basic sets* Φ. Polynomials of Σ which are not in Φ can be pseudo-divided by all polynomials in Φ. All non-zero pseudo-remainders can be adjoined to Σ and create a new polynomial set Σ', with new *basic set* Φ', and so on. Eventually, all polynomials of $\Sigma^{(i)}$ will either be in the *basic set*, or be algebraically dependent on it (their pseudo-remainders will be zero).

Ritt even mentioned the computational procedure for constructing remainders. Any differential polynomial G not in the ascending set $A = \{A_i\}$ was reduced, in several iterations, to a remainder R lower in the order of derivatives, and in the degree of leading variables:

$$G \prod_i S_i^{s_i} I_i^{t_i} = \sum_i \sum_j Q_{ij} A_i^{(j)} + R \tag{6.17}$$

The two symbols S_i and I_i are the 'separant' and the 'initial' of the forms in the ascending set:

$$A_i = I_i \left(\frac{d^{q_i} y_i}{dx^{q_i}} \right)^{r_i} + \cdots$$

$$S_i = \frac{\partial A_i}{\partial y_i}$$

Although reduction played a central role in Ritt's study, it was used as a guarantee that polynomials with required properties can be generated in a finite number of steps, rather than as an algorithm that would be actually performed. Ritt's final book *Differential Algebra* (Ritt 1950) devoted even more attention to abstract concepts such as ideals and bases, and to the structure of differential fields. Ritt's students, notably E.R. Kolchin and H.W. Raudenbausch, pursued the path opened up by Ritt into a direction more integrated with the development of abstract algebra. Ritt also changed his terminology – differential forms became 'differential polynomials', ascending sets 'chains', and basic sets 'characteristic sets'.

In this form, Ritt's work became widely known and used by analysts and algebraists alike. Its original algorithmic form was, however, exploited neither by Ritt himself, nor by his students, nor by other mathematicians prior to Wu Wen-Tsun.

The influence of Ritt's method on Wu was lasting, but developed only gradually. In Wu Wen-Tsun (1977a) he observed that 'the notion [of a privileged basis] is in reality due to Gröbner under the name of prime basis, cf. e.g. (Gröbner 1949) and cf. also (Ritt 1950) for the intimately related concept of characteristic sets.' But he neither used Ritt's techniques in the examples illustrating his method, nor based his proofs exclusively on Ritt. He tried to formulate the theoretical basis of his method – the decomposition into irreducible varieties – on factorization and calculation of the basis of the primary ideal. He only used parts of Ritt's framework for its theoretical justification.

Wu's first use of the full power of Ritt's method is visible in Wu Wen-Tsun (1978b), an attempt to expand his method to elementary differential geometry. Later, in Wu Wen-Tsun (1984b), he provided a full exposition of Ritt's theory of polynomials, ascending and basic sets, as well as the reducibility of algebraic varieties. Significantly, this time he went back to Ritt's earlier book from 1932 and reinstated its terminology (basic instead of characteristic sets, ascending sets instead of chains).[5] Perhaps this was a gesture towards the 'classical simplicity' and more algorithmic approach of earlier algebra that Wu wanted to revive in his work. In this respect, a passage he wrote about the history of invariant theory is interesting:

> A turning point in mathematical research in the West was invariant theory. Before that, mathematics had been constructive, and it was a widespread requirement that it must be. In order to prove existence, one had also to give a method of construction at the same time, which was of course a constraint. (…) One basic problem was whether it is possible to find a finite number of invariants such that all the other can be expressed as their polynomials. At that time, research was done by laborious calculation, and Gordan completely solved the case for two variables. The problem became very difficult for three or more variables, but Hilbert quite elegantly proved the existence of a finite basis, which it was hard to accept from the constructive outlook current at the time.
>
> Invariant research has submerged because of its complexity, but mathematics cannot avoid this important question, as Dieudonné has said: 'Invariant theory is as a phoenix, always rising from its ashes.' (…) Most recently, the *Bulletin of the American Mathematical Society* carried a long article introducing nineteenth century invariant theory, suggesting many new problems. Some related nineteenth century books are again being reprinted, this is also a sign of constructive mathematics reasserting itself.
>
> (Wu Wen-Tsun 1985a: 335)

Here Wu considered Hilbert's proof of the finite basis theorem as the start of non-constructive mathematics and, consequently, he wanted to exorcise these methods and essentially non-constructive, non-finitist concepts (such as the concept of prime ideals and their bases) from his theory altogether. He found the earlier version of Ritt's theory more helpful.

Wu's interest in Ritt's method brought its neglected algorithmic quality to the foreground. Wu was probably motivated in his preference for Ritt's method not only by its practical advantages in comparison to the Gröbner bases, for example, but also, equally strongly, by the consistency of Ritt's approach with Wu's post-1970s convictions about the nature of mathematics. Wu came to believe that mathematics needed to become more mechanized and to focus on equations and algorithms for their solutions. He felt let down by more recent algebra and algebraic geometry in this quest for mechanization, and turned eagerly, as he had several times in his life, to the mathematics of the 1930s.

Wu's method and the limitations of Euclidean proof

One important aspect of Wu's method is the fact that it involves pseudo-division, rather than proper division. This means that the conclusion and then the subsequent pseudo-remainders have to be multiplied by initials of the premises, called non-degeneracy polynomials D_a. These polynomials, produced automatically by the elimination algorithm, have a geometric significance. Since they have to be non-zero, they introduce additional restrictions on parameters of the theorem. This corresponds to the fact that geometric theorems are almost never true in the strict sense, i.e. in all possible configurations of objects, but only in generic situations. Borderline configurations where the theorem might lose its meaning are defined by implicit non-degeneracy assumptions. For example, lines which are supposed to intersect by the premises should not be identical, points forming a triangle should not lie on the same line, and so forth.

The final step of Wu's method, interpretation, involves precisely considerations of the geometric meaning of the non-degeneracy conditions. Wu Wen-Tsun emphasized that his method always accurately produces these non-degeneracy conditions, which is its major advantage over the traditional proof. Geometric theorems proved mechanically were elevated to a higher standard of rigour, according to Wu. He noted that traditionally, any theorem should be investigated and proved separately for every degenerate case. But this is in fact impossible:

> Now to prove theorems in the usual Euclidean fashion, one should incessantly make [recourse] to previously proved theorems. (…) One should, each time when these theorems are to be applied, verify whether [their] non-degeneracy conditions are observed or not. One should consider different cases to [dispose] of one by one each of these degeneracy situations. (…) In fact, even for a theorem of moderate complexity, it would be quite impossible to take care of all these non-degenerate cases [occurring] in the known theorems to be applied.
>
> (Wu Wen-Tsun 1982b: 126)

He showed that his procedure generated the subsidiary non-degeneracy conditions automatically, and thus the theorems are not only proved, but made more precise and explicit by the procedure:

What is important for us is that the degeneracy conditions which may cause the fallacies of theorems [to] present [themselves] *automatically* during a mechanical procedure and may be treated alternatively and systematically also in a mechanical way, which is actually impossible for the usual Euclidean fashion proofs.

(Wu Wen-Tsun 1982b: 136)

Wu Wen-Tsun's attitude to the Euclidean proof was critical but, in the end, he regarded the Euclidean ideal of complete certainty as his own. Similarly, although he called for more weight to be given to the mechanical and algorithmic tendencies in mathematics as opposed to axiomatization, his work on the mechanization of theorem-proving actually engages quite thoroughly with the axiomatic foundations of various geometries, to which his method could be applied (the necessary condition is that the axioms make the geometry isomorphic to a field and thus allow easy computations). In all these cases, Wu acted as a committed mathematician accepting the intellectual ideals of his discipline, even though he tried to distance himself from them. He demonstrated a complicated, almost split, identity: at the same time an insider to his trade, and a permanent rebel inclined to creating mathematical countercultures. In the final chapter, I will try to show how these two tendencies could be linked.

Notes

1 For a comprehensive historical overview of different approaches to mechanization of mathematics, see Beeson (2004).
2 The division was therefore regarded as unsatisfactory in the longer perspective, and the three institutes were combined again in 1998 into the current Academy of Mathematics and Systems Science.
3 For geometric theorem-proving, any method that involves polynomial elimination is likely to be quite inefficient compared to coordinate substitution (parametric equations; Doron Zeilberger, personal communication).
4 See the modern critical edition (Zhu Shijie [1303] 2007) and the English translation (Hoe 2008).
5 Chou Shang-Ching (1988), however, used a mixed terminology of ascending and characteristic sets, and this is what eventually became standard.

7 Saving the nation with mathematics and its history

How will history evaluate Wu Wen-Tsun's dual turn to 'mechanization' and to traditional Chinese mathematics? Did he succeed in reinvigorating independent mathematical research, and thus help the Chinese nation? Or was it only a remarkable reinvigoration of his own career, without lasting consequences for Chinese mathematics as a whole? It is too early to answer these questions, and it would require a thorough assessment of current trends in Chinese mathematics, an exercise beyond the scope of this book and the competence of its author. In this final chapter, I shall instead summarize what the dual turn has brought to Wu Wen-Tsun personally, and discuss to what extent it is workable and acceptable as a model for the Chinese mathematical community.

The analysis will proceed in four steps. The first section compares the influence of his papers and other publications from different periods of his life. Insofar as quantitative data can tell any story at all, they strongly suggest that Wu's method of mechanization is the decisive achievement which has made his name.

The second section returns to the realm of culture and symbolic meanings, and analyzes Wu Wen-Tsun's attitudes to modern and traditional mathematics using rather old-fashioned theories of cultural nationalism. It will be seen that Wu Wen-Tsun's historicism is indeed an extension into the realm of mathematics of processes that have often been observed and studied in other areas of modern nation-building. But since such cultural attachments of mathematics continue to strike many Western and Chinese mathematicians as strange and even dangerous, it is useful to contrast in the third section Wu's strategies of mathematical nation-building with those of his famous compatriots. This comparison suggests that even though other mathematicians might appear more universalist than Wu, most do share the primary goal of building a strong Chinese mathematics and think about different paths from the most mainstream Western mathematics.

The final section, indeed an epilogue, will show Wu Wen-Tsun's rise in the power structures of China's science and politics. His personal success is a strong indication that although not everyone shares his attitude towards traditional Chinese mathematics, his broader goals are completely in line with the visions of the Chinese mathematical community and political leadership.

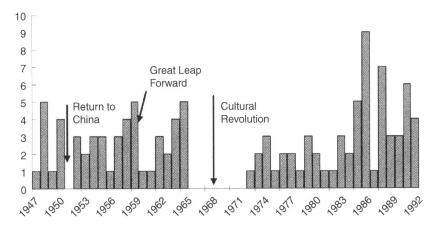

Figure 7.1 Number of Wu Wen-Tsun's publications per year, 1947–93

Note: Publications are shown in the year of their first appearance in whatever language. Preprints followed by journal publications shortly afterwards are not listed separately.

Wu Wen-Tsun's research and international recognition

Although Wu Wen-Tsun has had a lifetime record of productivity and mathematical creativity (Figure 7.1), the years after 1985 have been the most successful in terms of both the number of publications and the number of highly cited publications. He published at least three papers per year in the late 1980s and through most of the 1990s, although many appeared only in the non-peer-reviewed *Mathematical Mechanization Research Preprints*, created by his 'Mathematics Mechanization Research Centre' in 1990.

Wu's pre-1985 activity peaked three times: first in France, with his classic results on characteristic classes; then at the height of the Great Leap Forward, when publication of earlier results on realization of polytopes coincided with new investigations of game theory; and for the third time in 1963–64, when he revisited the theory of immersion and at the same time studied singularities and algebraic geometry. On the other hand, there was a marked post-GLF slump in 1960–61, and of course the interruption by the Cultural Revolution.

When we look at citation counts, different databases give slightly different results (Table 7.1, Figures 7.2–7.5). MathSciNet is true to its most mathematical nature and accords the top position to Wu's 1965 book on imbedding, and several other high ranks to his French and early Chinese work in algebraic topology. Google Scholar, on the other hand, ranks the recent mechanization books and articles much higher, with the top publication scoring more citations (436) than MathSciNet records in total for all of Wu's 126 publications listed there (367). The smaller databases are less useful for comparison,

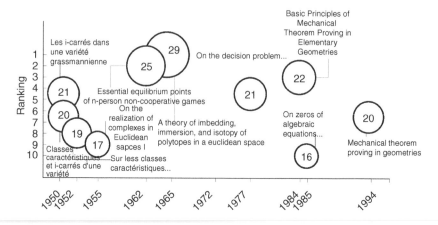

Figure 7.2 Citation counts of Wu's publications in *MathSciNet*

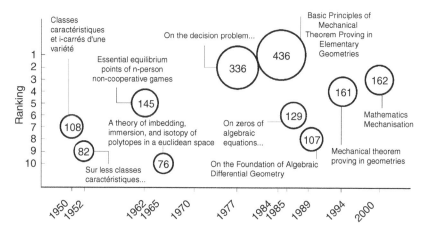

Figure 7.3 Citation counts of Wu's publications in *Google Scholar*

because they do not register citations of books. Even with this limitation, they corroborate the feeling that Wu's post-1978 work was, and especially is today, much more influential than his early papers. A slightly unexpected result is the consistently high place of Wu Wen-Tsun's work on non-cooperative games together with his student Jiang Jiahe, which is the highest pre-1978 result in all databases apart from MathSciNet, and even there it ranks second.

The 25 highest-impact publications listed in the four 'Top Ten' tables cover almost Wu's entire career. In terms of breadth and citation counts, the algebraic-topological stage of Wu's career was slightly less prominent, but still followed with interest by many outstanding foreign colleagues (Stone 1961; MacLane 1980). Wu's results were directly related to the work of at least five

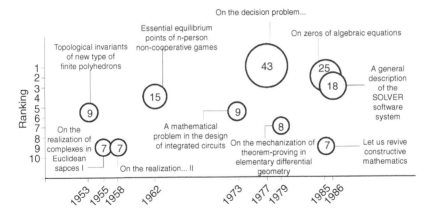

Figure 7.4 Citation counts of Wu's publications in the *Web of Science* and *Scopus* (dotted circle)

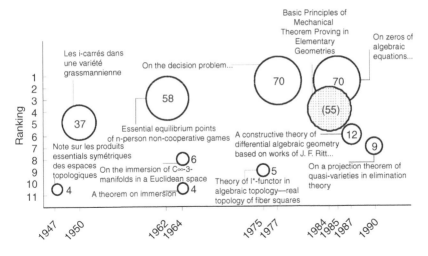

Figure 7.5 Citation counts of Wu's publications in the *China National Knowledge Infrastructure*

Fields medallists from the years 1954–66: J.-P. Serre, R. Thom, J. Milnor, M. Atiyah and S. Smale. Indirect connections can also be found to a sixth (S. Novikov, FM 1970). The fact that almost all his publications were reviewed by *Mathematical Reviews*, including those written in Chinese and published in less prominent journals, also attests to his international renown. But Wu's publications exclusively in Chinese were in fact quite limited in number, and mostly concentrated in the years of isolation during the Cultural Revolution period, and in the history of mathematics (Table 7.2). The near absence of French-language publications after Wu's return to China is also remarkable,

Table 7.1 Wu's ten most cited publications with citation counts in different databases

Citations in MathSciNet	Citations in Google Scholar	Citations in Web of Science or Scopus	Citations in CNKI
A theory of imbedding…(1965a) 29	**'Basic principles…' (1984a)** 436	**'On the decision…' (1977b, 1978d)** 70	**'On the decision…' (1977b)** 43
'Essential equilibrium…' (w. Jiang Jiahe 1962) 25	**'On the decision…' (1977b, 1978d)** 336	**'On zeros…' (1985b, 1986a)** 70	**'On zeros…' (1985b)** 25
'Basic principles…' (1984a) 22	**Mathematics Mechanization (2000)** 162	'Essential equilibrium…' (w. Jiang Jiahe 1962) 58	**'A general description of the SOLVER…' (1986d)** 18
'Les i-carrés…' (1950b) 21	**Mechanical theorem… (w. Wang Dongming & Jin Xiaofan 1994)** 161	**'Basic principles…' (1984a)** <u>55</u>	'Essential equilibrium…' (w. Jiang Jiahe 1962) 15
'On the decision…' (1977b, 1978d) 21	'Essential equilibrium…' (w. Jiang Jiahe 1962) 145	'Les i-carrés…' (1950b) 37	'Topological invariants…' (1953b) 9
'Classes caractéristiques…' (1950a) 20	**'On zeros…' (1985b, 1986a)** 129	**'A constructive theory…' (1987g)** 12	'A mathematical problem…' (1973) 9
Mechanical theorem proving… (w. Wang Dongming & Jin Xiaofan 1994) 20	'Classes caractéristiques…' (1950a) 108	**'On a projection theorem…' (1990b)** 9	**'On the mechanization…' (1979a)** 8
'Sur les classes…' (w. Reeb 1952) 19	**'On the foundation…' (1989b)** 107	'On the immersion…' (1964b) 6	'On the realization…I' (1955d) 7
'On the realization…III' (1958b) 17	'Sur les classes…' (w. Reeb 1952) 82	'Theory of $I*$-functor…' (1975c) 5	'On the realization…II' (1957b) 7
'On zeros…' (1985b, 1986a) 16	A theory of imbedding… (1965a) 76	'Note sur les produits…' (1947) 4	**'Let us revive…' (1985a)** 7
		'A theorem on immersion' (1964a) 4	

Note: Citations of Chinese originals and translations are aggregated. Bold titles indicate post-1978 work related to mechanization.

Table 7.2 Wu Wen-Tsun's mathematical publications by topic, language and period

Topic	Publication language					Time span
	Fr	En	Ch+En	Ch	All	
Algebraic topology and geometry	12	16	14	5	47	1947–65
Applied and transitional mathematics		7	4	6	17	1959–87
Mechanized and constructivist mathematics		60+	11	16+	87+	1977–2009?

as well as the number of English-only articles in the 1960s. This highlights his determination to safeguard his international position in that period, and perhaps also incentives from the leadership of the Academy of Sciences to face the Western world more proactively.

I have clustered Wu's publications in three groups rather than just two (algebraic topology and mechanization), because there is an interesting 'transitional' zone between them. This includes research in pure mathematics (I^*-functor), as well as applied, but in all cases pursued with clearly stated ideological aims, which was not the case in the 'pure' period. Common to these disparate fields, often abandoned after relatively short periods, is Wu's desire to infuse his mathematical results with some kind of wider social or political relevance. They can thus be seen as initial attempts to break away from the kind of modern mathematics in which Wu had been trained, and whose leading protagonists he was competing with, for the larger part of the 1950s and 60s.

Nationalism as a creative and disturbing reaction

Wu Wen-Tsun's turn to history has several positive and creative aspects. He has brought new perspectives to the history of Chinese mathematics, helped to establish and promote a more nuanced and sophisticated view of the strengths of Chinese mathematics in comparison with mathematics born of Greek geometry, and stimulated an interest in the history of mathematics among a number of Chinese professional mathematicians. His direct borrowing from Chinese mathematical techniques, albeit in a context dominated by modern mathematical theory, is a powerful demonstration that the history of older mathematics can be relevant to contemporary research (in the frequently repeated phrase, *gu wei jin yong*).

But along with all these positive achievements, Wu Wen-Tsun also claims that there is an essentially Chinese mathematical style, and that Chinese mathematicians have a patriotic duty to study it and build upon it. This makes many people worried. Is mathematics not a universal human endeavour, where the criteria of what to study and use should be common to all mathematicians?

Could parochial adoration of Chinese mathematical style lead the next generation of Chinese astray from committed study of the most difficult, most cutting-edge mathematics regardless of its origin? Given the sorrowful history of mental isolation and repeated attacks against 'excessive Westernization' in China, one can understand the concern among some that patriotism should not interfere with choices of scientific style, and obscure rational choices with what they see as emotional attachments to traditions – traditions that have long died out and are suddenly restored by fully Westernized people like Wu Wen-Tsun. Moreover, given the twentieth-century ravages of nationalism, racism and chauvinism in various forms, anyone riding the wave of national sentiments is regarded with suspicion.

We might easily dismiss the nationally hubristic component of Wu Wen-Tsun's historicism as partly showmanship intended to secure precious resources in the post-1978 competitive social climate, partly a lingering atavism of the Cultural Revolution, or perhaps an unfortunate mental scar resulting from the suffering of those years. Many lamentable stories have been mentioned on the preceding pages: Hua Loo-Keng's retreat to the popularization of operations research among peasants and workers, Chen Jingrun's solitary work under fear of persecution, or the actual persecution of Wang Yuan, Zhang Jingzhong and many others. Wu Wen-Tsun's historicism could be explained as simply an internalized self-preservation strategy adopted under political pressure and retained for complex psychological reasons.

But this would be a mistake similar to the long-held view of early-twentieth-century Chinese cultural conservatives as poor souls lost in the modern age and longing for certainties of China's imperial past (or cynically catering to such longing among Chinese masses). As has been quite decisively shown in recent scholarship (Duara 1995; Fung 2010), cultural conservatism is simply an alternative response to modernity sharing many of the same values and goals as radical modernizing ideologies, but sceptical of the means through which these values and goals are to be achieved. In fact, most Chinese mathematicians are openly nationalist, have contemplated strategic visions of successfully competing with the West, and even speculated about a specifically Chinese path to modern mathematics. Wu's strain of historicist nationalism is simply an attempt to mobilize cultural resources for the same aim.

Cultural nationalism was critically analyzed by Ernest Gellner as a logical but intellectually untenable response to uneven modernization (Gellner 1964: 147–78). Modern society and economics favour universalism, but the uneven progress of modernization gives some cultures early advantage and bars those who are not part of them from fully sharing the fruits of modernity. Elites of the disadvantaged peoples build rival 'high cultures' on the patchwork of whatever bits and pieces of traditions they can find in their homeland, and protect both these traditions and their own social status from the encroachment of universalist empires. Although Gellner ridiculed with great gusto the often disingenuous pathos and myth-building excesses of cultural nationalists (Gellner 1983: 48–62), he speculated that they actually prevent the tendency of large modern universalist states to create de facto racist hierarchies (Gellner 1994; Nairn 1997).

It is possible, as I have done elsewhere (Hudecek 2012), to view Wu's construction of Chinese traditional mathematics as precisely such an exercise in nation-building in order to preserve a breathing space for independent, sovereign research agendas. This would resonate with his reinterpretation of the 'Bourbaki myth' (Corry 2004) as an expression of French national spirit (see Chapter 6). Although the Bourbaki mathematicians indeed aspired to overcome through their activities a perceived decline in French mathematics, there is no indication that they were cultural nationalists.[1] Wu's nationalist interpretation of Bourbaki only emerged in 1985, when he was engaged in his own 'nation building' – his earlier article about Bourbaki, written shortly after returning from France, did not even mention their discontent with the state of French mathematics, let alone nationalism (Wu Wen-Tsun 1951). This indicates that in 1985 he was exploiting the legend of Bourbaki, and the undeniable fact that Bourbaki did in fact revive French mathematics, rather than remembering his own earlier experiences.

But a better way to understand Wu's interest in Chinese mathematical history might be to take both his cultural nationalism and his critical attitude to some trends in modern mathematics more seriously and observe what he was actually trying to achieve. Wu's historical writings were not only patriotic expressions of justified outrage at the downplaying of Chinese contributions by Western historians of mathematics. They also argued that traditional Chinese mathematics could be relevant for achieving mathematical independence and that it embodies a progressive mathematical philosophy. Wu was arguing for a national revival, and at the same time justifying his own work, appealing to his colleagues' sense of responsibility:

> Question Q: In the coming era of computer age, what [should] the mathematicians do? In particular, what [should] the *Chinese* mathematicians do?
> (Wu Wen-Tsun 2002: 4)

Wu Wen-Tsun has been celebrated for his methodology of research into the history of ancient Chinese mathematics. But ultimately, the goals of Wu's historiography were presentist – to establish ancient Chinese mathematics' close contacts with fundamental mathematical realities as understood by him, a twentieth-century mathematician. The apparently anti-presentist methodology was only a certain precautionary device, rather than a belief that ancient mathematics should be investigated on terms truly its own.

This paradox can be explained by looking at Wu's experience as a mathematician. He frequently revisited older literature and thought about its methods and viewpoints that had subsequently been lost or diverted: non-homotopy-type invariants and van Kampen's method in the 1950s; the use of the Smith–Richardson calculus for an intrinsic definition of Steenrod squares in the same period; the use of van der Waerden's theory of generic points for his work on algebraic geometry in the 1960s; and the use of J.F. Ritt's theory, in its older version, in the method of mechanization. Wu approached these earlier works attentively and with an open mind to find ideas missing in current mathematics. At the same time, he frequently translated their language into more modern terms, because he had no doubts that the underlying objects were identical, and he strove for contemporary relevance.

Wu brought this way of reading old mathematics over to his study of ancient Chinese mathematics. On the one hand, the unfamiliarity of these texts and their language suggested greater caution, if no misunderstandings were to be introduced and no subtle ideas missed. On the other hand, historical aspects could be discarded if they did not, in Wu's view, contribute anything potentially relevant and interesting from the perspective of modern mathematics (for instance, he never devoted more than a brief mention to the counting rods with which ancient and medieval Chinese mathematicians performed their calculations).

Wu saw Chinese mathematics in evolutionary terms, and wanted to capture and use its core strengths. He can be characterized as a reformist within the classic triplet of neo-traditionalist, assimiliationist and reformist 'routes to ethnic historicism' (Smith 1981: 96–107). He abandoned the assimiliationism dominant among China's intellectuals, the 'embrace with an almost messianic fervour' of modern society, which only uses history for clues to establishing a local version of the 'rational, progressive and scientific state' (Smith 1981: 99–102). Many of his colleagues could be listed under this position, as we shall presently see. As a reformist, on the other hand, Wu wanted to find a compromise between modern overly abstract mathematics and traditional calculations close to applications, to 'salvage the true, the underlying, the pure' form (Smith 1981: 103) of Chinese mathematical tradition. This would provide Chinese mathematics with a source of vitality – the exploitation of its best traditions – and at the same time stay close to what mathematicians should in his view be doing – calculating and creating methods to calculate more effectively.

Historicism is a consciously rigorous and 'scientific' position, which aims to present rational arguments for the selection of the most viable and promising features of the (constructed) tradition. This scientific rigour is achieved through historical accuracy. On the other hand, the reformist position also places conscious limits on reason, subordinating it in certain situations to the power of the emotional attachment to tradition. The rigorous analysis of Wu's most scholarly historical articles is only one facet of his approach to history. In other cases, he talks about the present and the future in historical parallels intended as an illustration and spur to action (Wu Wen-Tsun 1987b), as with the analogy between the mechanization of manual and mental labour.

Indeed, reformism is a permanent grand compromise between cold rationality and fervent feelings towards the tradition. Wu decided to live such a compromise by returning to China, despite knowing that he could stay abroad and work under better conditions. More than many of his colleagues, Wu eagerly sought a compromise that would give his return to China its ultimate meaning. Mathematics based – however loosely – on an understanding of ancient Chinese tradition satisfied this desire.

Wu Wen-Tsun has said that he 'could not have gained' the understanding of ancient Chinese mathematics without his experience in China.[2] This might appear puzzling – the texts have always been available abroad, there were also scholars studying them, and Wu would likely have time to devote to such hobbies in older age.[3] Maybe the point Wu wanted to make was really about his choice: had he stayed in France or gone to, for example, the United States, he would have abandoned the compromise once and for all. Studying ancient Chinese mathematics would thus not only be more difficult, but also far less meaningful.

Wu Wen-Tsun's search for a specifically Chinese mathematics is thus intimately connected with his rationalization of the decision to return to China in 1951. Wu's apparently resolute verdict that it was right to return, because only this enabled him to understand the qualities of ancient Chinese mathematics, shows that he wants both China and world mathematics to benefit from the opportunities he has given up. Wu's cultural nationalism is thus built as much on his lifelong patriotism as on the sacrifices he had to make because of it.

Nationalism and universalism in modern mathematics

Nationalism oftentimes strikes utilitarian compromises with other agendas and programmes. The question then arises – how difficult is the compromise with mathematics? Is mathematics not a universal, supranational endeavour?

Indeed it is, in some sense – the objects it investigates are universally comprehensible, despite subtle variations in concepts, their paradigmatic instances, and theoretical approaches between national schools that are often overlooked but exist even in modern mathematics.[4] From the sociological perspective, mathematics was for the most part of its history a very rare occupation. Mathematicians lived far apart, and were happy to find anyone with the same interests. This created an influential image, preached by mathematicians such as David Hilbert (1862–1943), of the 'universal brotherhood of mathematicians' (Rowe 2003: 535). Hilbert allegedly said 'Mathematics knows no races... For mathematics, the whole cultural world is a single country'.[5] Despite the spurious nature of this quotation and its narrow original context, celebrating the renewed access of German mathematicians to international congresses, it conveys very well the feeling that common interests in mathematical objects should come before national rivalries.

In fact, international communication can be and has often been accompanied by competitive nationalist tendencies, even in mathematics. The era of Hilbert's alleged remark, between the two World Wars, was a typical example. Even the ostensibly internationalist congresses were often merely displays of national competition and achievements of particular schools (Segal 2002).

It is clearly possible to accept the international nature of mathematics in practice, without subscribing to mathematical supranationalism as a principle.

Mathematicians of course have to share, communicate and adopt techniques, symbols, research agendas, and so forth. Wu Wen-Tsun has done so throughout his career, as he combined French, Swiss, German, Russian and American influences in his work, and used all available opportunities to present his results to international audiences. He was obviously upset when these communications were blocked, and after 1978 eagerly used China's opening up to travel to conferences and foreign research centres, and invite famous foreign mathematicians to China. On a purely pragmatic level, he has accepted the international nature of mathematics without reservations. But since the Cultural Revolution, he has also seen international exchanges in nationalist terms, as communication between different nations – equal, mutually respecting, but separate and competing. And he has repeatedly claimed that Chinese mathematicians have specific responsibilities by virtue of being Chinese.

A competitive understanding of international mathematical exchanges, and the (intuitive) patriotic or (conscious) nationalist position in them, can be documented for many famous Chinese mathematicians. It was clearly so with Hua Loo-Keng, but Wu's teacher S.S. Chern is perhaps an even better example. Having spent 50 years in the United States, after he declined to return to China in the 1950s,[6] he nevertheless devoted, even as a naturalized US citizen, much of his time and energy in the last 32 years of his life to the strengthening of Chinese mathematics, especially through his support for the Institute of Mathematics at Nankai University. He is reported to have said: 'Chinese mathematics must be on the same level as its Western counterpart, though not necessarily bending its efforts in the same direction' (Hitchin 2006: 509).

Chern showed his vision for the development of Chinese mathematics during his short tenure of slightly more than two years (1946–48) as acting head of the Preparatory Office and later acting director of the Institute of Mathematics of Academia Sinica in Shanghai. He selected a dozen young assistants to be trained in the branch that he considered most promising for the future development of mathematics, algebraic topology, for him a new tool for the solution of open problems in his own field (differential geometry) as well as many others. According to his later recollections, he was inspired by the example of interwar Poland, which first gained prominence in functional analysis and set theory and gradually expanded its mathematics in other branches as well (Chern Shiing-Shen 1988).

Chern did not share Wu's enthusiasm for ancient Chinese mathematics,[7] and was especially allergic to narrow-minded utilitarianism limiting mathematical research (Tian Miao 2000). He once wrote a review article on the entire history of Chinese mathematics, where he suggested that the applied character of Chinese traditional mathematics should inspire more attention to applications both among mathematicians and among non-mathematicians encountering possibly mathematical problems in their work. But he left this task for 'some people' in the discipline. His article was also notable for refusing to recognize any essentialist division between ancient and modern Chinese mathematics. He instead followed the rise and fall of the 'research spirit' of Chinese mathematics as the defining quality of mathematical culture at any given time (Chern Shiing-shen 1941).

Chern wanted to develop promising but relatively unfashionable areas, where it is easy to derive new results and assume leadership. This was Chern's explanation for his own success in differential geometry in the 1940s (Hu Sen 2007), and a principle which Wu tried to emulate throughout his mathematical career. Chern was, at the same time, an extremely cosmopolitan mathematician (Yau, *et al.* 2011).

Another famous example of a nationalist Chinese mathematician might be Chern's student Yau Shing-Tung (S.T. Yau) (born 1949, Fields Medal 1982), who caused a scandal in 2006 by promoting the work of his two students on the Poincaré conjecture as a 'complete proof' (Nasar and Gruber 2006). Yau invoked the nationalist theme explicitly when he said 'Chinese mathematicians should have every reason to be proud of such a big success in completely solving the puzzle.' In fact the work was of lesser significance, and Yau was then suspected of trying to steal the credit for solving the Poincaré conjecture from the Russian Grigory Perelman (born 1966), who had been awarded (but declined) the Fields Medal for his achievement in 2006 (Alexander 2010: 168–70). The *New Yorker* article which revealed this story also speculated that Yau wanted to cement his status as the leading student of S.S. Chern, and thus the leading figure of Chinese mathematics. This shows the self-reinforcing effects of nationalism when it becomes a tool for obtaining power and status.

A rich international experience, then, does not diminish patriotism. On the contrary, it might make nationalist sentiments more acute and more sophisticated, even adopting a historicist route of 'assimilationism', as famous Chinese radicals (Hu Shi, 1891–1962, among others) had done in the early twentieth century. Admirers of foreign cultures can be turned into nationalists as they are denied full access to the foreign society, observe its unresolved problems, and realize that the difference in development between their home and the country they want to emulate is purely historical and could be reversed. They might also realize how strong and effective nationalism is over there, as Wu Wen-Tsun did when he saw Bourbaki as the product of the French national spirit.

Wu Wen-Tsun's social rise since 1978 and his legacy

Let us at the very end of this text return to the facts of history: Wu Wen-Tsun's rise to a respected and powerful figure since the beginning of the Reform and Opening in 1978. In these years, Wu gradually became active in both academic and national politics and in the setting of China's science policy. As early as 1978, he was appointed to the Standing Committee of the Chinese Political Consultative Conference (SCCPCC), and served on it for the next 20 years. He was also chosen by Guan Zhaozhi (who died in 1982) to act as his deputy director at the Institute of Systems Science (ISSCAS), and has been its honorary director since 1984. He also became a member of the executive committee of the National Science and Technology Fund, the highest decision-making body for allocation of research funding, and served as head of the Academic Department of Mathematics, Physics and Chemistry of CAS in 1992–4.

In 1980, Wu was elected to the Vice-Presidency and in 1984 to the Presidency of the Chinese Mathematical Society (CMS). In this capacity, Wu negotiated a membership settlement with the International Mathematical Union (IMU) for CMS in 1986. The CMS had, since the 1950s, declined direct engagement with the IMU because of the contested status of Taiwan (represented in IMU as Republic of China). After intensive negotiations in the 1980s, IMU followed a formula established by the International Union of Pure and Applied Chemistry, whereby Taiwanese members were officially included in the Chinese section, but identified separately as China (Taipei). The negotiations with IMU were complicated by the need to change its constitution to allow for this special situation (Lehto 1998: 242–50). The integration of CMS into the IMU culminated with the holding of the International Congress of Mathematicians in Beijing in 2002, which Wu opened as the Honorary Chair of the Local Organizing Committee (Hu Zuoxuan and Shi He 2002: 207).

Wu's success was also recognized through various prizes and prestige positions – membership of the Third World Academy of Sciences in 1991, the Herbrand Prize for Automated Reasoning in 1997, the first Highest National Science and Technology Award in 2001, and the Shaw Prize for Mathematical Sciences in 2006, with the 1974 Fields medallist David Mumford (Hu Zuoxuan and Shi He 2002: 170–85; The Shaw Prize Secretariat 2006).

The Chinese Highest National Science and Technology Award of 2001 generated a new wave of interest in Wu Wen-Tsun within China, and he has since become something of a celebrity, although not quite on a par with more famous old scientists (including his teacher S.S. Chern). Wu has never actively sought fame, and he became increasingly reluctant to give interviews and attend public meetings as his age advanced. In 2008, his health deteriorated, and although he has somewhat recovered, he now rarely leaves home, and spends less time with mathematics (Zhang Xianfeng 2011).

Wu also received some recognition for his historical work. He was invited to present a survey of recent studies of the history of Chinese mathematics at the International Congress of Mathematicians in Berkeley (Wu Wen-Tsun 1986b), where he avoided his usual nationalist arguments but kept his general ideas about the main features of ancient Chinese mathematics.

Part of Wu's status in the historical community comes from his active support for the history of Chinese mathematics. He set up a 'Mathematics and Astronomy Silk Road Fund', using 1 million yuan, twenty per cent of the cash prize he received in 2001. The fund finances investigation of intercultural transmission of mathematics, with the hope that new evidence of Chinese influence on other Asian countries would eventually be found (Li Wenlin 2010). Chinese historians of mathematics have often invited Wu to act as (honorary) editor of conference volumes and collective works, including the *Great Series on the History of Chinese Mathematics* (*Zhongguo shuxue shi da xi* 中国数学史大系, 1998–2000). The prefaces to these works have, in turn, been an important dissemination channel for Wu Wen-Tsun's ideas and the story of his inspiration from traditional Chinese mathematics.

Wu Wen-Tsun's early success in mechanical theorem-proving did not lead (or perhaps has not yet led) to a permanent revival of ancient Chinese mathematics. In this sense, he did not discard the 'Needham Question'. But he could serve as an example of why the 'Needham Question' really should not matter so much: the positive aspects of the tradition should be studied regardless of its 'failures' and inefficiencies in other respects or other time periods. Indeed, Wu never wrote more than half a sentence about the decline of ancient Chinese mathematics. It seems that for him, this was simply a part of non-mathematical history and could not fundamentally change the evaluation of the earlier mathematical content. He has always only been interested in its strengths and contributions. This bias and selectivity are simply an expression of his belief that to accept and study tradition was a way of giving meaning to his Chinese identity. In the end, the point was not to revive ancient Chinese mathematics, but to revive Chinese mathematics *tout court*. Although he stands apart from most other Chinese mathematicians by looking to the past, he shares the same goal with them of opening the future to a revived Chinese mathematics, a mathematics serving the symbolic and practical needs of the society that nurtures it.

Notes

1 Cf., however, Mathias (1992), who speculated that Bourbaki ignored the development of logic for nationalistic reasons. The motivation to overturn a decline in French mathematics was mentioned by Weil (1992).

2 From Li Xiangdong and Zhang Tao (2006: 37:30–38:30), also quoted in Ke Linjuan (2009: 160–61). Wu made the same argument, without any prompting, in the interview with the author on 10 July 2010.

3 There are several mathematicians and physicists of Chinese origin active in other countries who, apart from their successful primary careers, also devote some of their time to the history of Chinese science, often with an explicit Chinese patriotic emphasis. Many of J. Needham's collaborators can serve as examples (e.g. Ho Peng-Yoke, H.T. Huang). An example particularly close to Wu Wen-Tsun is Joseph Chen Cheng-Yih 程贞一 (born 1933), a professor of nuclear physics at the University of California, San Diego, who has also devoted attention to the study of ancient Chinese algorithms. His relationship with Wu was strengthened by the fact that one of Wu's daughters has lived in San Diego since 1995 (Chen 2010).

4 See, for example, the differences between British and German invariant theory in the nineteenth century (Parshall 1989).

5 Reid (1970: 188) quotes this remark as the conclusion of Hilbert's informal speech upon arrival to the International Congress of Mathematicians in Bologna in 1928, the first since the First World War to which German mathematicians were invited. The remark also circulates in the version 'Mathematics knows no races *or geographic boundaries*', apparently going back to the collection of anecdotes (Eves 1972: 136). Neither Reid nor Eves reveal the source of this quotation; Eves acknowledges that

'many of the stories we tell here about Hilbert are also told, sometimes in a different version', in Reid (1970). Reid's book was mainly based on interviews, and it is conceivable that one of her interviewees told Eves the story too; but it seems more plausible that he simply copied the quotation (which runs to three paragraphs) from Reid and slightly changed it to avoid outright plagiarism. (Eves also erroneously talks about the Second ICM, when it was in fact the sixth.) Curiously, some scholars quote Hilbert in the Eves version but attribute it to Reid, even with page numbers, or with an obviously reverse-translated German 'original' (Thiele 2005: 254). Neither the otherwise very detailed six-volume proceedings from the congress (*Atti* 1929), nor contemporary reports (Gingrich 1929; Tonelli 1929; Vetter 1929), make any mention of a public speech where Hilbert would have talked about the internationalism of mathematics.

6 Chern's younger colleague Xu Lizhi speculated that Chern's reluctance to return was caused as much by material and political factors (Chern had shown relatively 'rightist' attitudes before 1949) as by the fact that the top job in Chinese mathematics had already been taken by Hua Loo-Keng (Xu Lizhi, *et al.* 2009: 230–31).

7 Chern's differences with Wu have been suggested by some informants in China, but are not so easy to document. Chern has only repeatedly expressed his belief that ancient Chinese mathematics, despite its achievements, was fundamentally hampered by its lack of rigorous deductions, and the achievements of Song and Yuan algebra were made 'despite the inefficient method' (Chern Shiing-Shen 2001).

Bibliography

Abhyankar, S. (1976) 'Historical ramblings in algebraic geometry and related algebra', *The American Mathematical Monthly*, 83 (6): 409–48. Online. Available HTTP: <http://www.jstor.org/stable/2318338> (accessed 25 June 2010).

Adams, J.F. (1961) 'On formulae of Thom and Wu', *Proceedings of the London Mathematical Society*, 11 (1): 741–52.

Alexander, A. (2010) *Duel at Dawn: Heroes, Martyrs, and the Rise of Modern Mathematics*, Cambridge, MA: Harvard University Press.

Alitto, G. [1979] (1986) *The Last Confucian: Liang Shu-ming and the Chinese Dilemma of Modernity*, 2nd edn, Berkeley: University of California Press.

Amelung, I. (2003) 'Die Vier großen Erfindungen: Selbstzweifel und Selbstbestätigung in der chinesischen Wissenschafts- und Technikgeschichtsschreibung', in I. Amelung, *et al.* (eds) *Selbstbehauptungsdiskurse in Asien: China – Japan – Korea*, München: Judicium.

Andreas, J. (2009) *Rise of the Red Engineers: The Cultural Revolution and the Origins of China's New Class*, Stanford: Stanford University Press.

Arkowitz, M. (1989) 'Wu, Wen-tsün (PRC-ASBJ-S) Rational homotopy type. A constructive study via the theory of the I^*-measure', *Mathematical Reviews Online*: MR903452. Online. Available HTTP: <http://www.ams.org/mathscinet/> (accessed 23 November 2013).

Atiyah, M.F. and Hirzebruch, F. (1959) 'Riemann-Roch theorem for differentiable manifolds', *Bulletin of the American Mathematical Society*, 65 (4): 276–81. Online. Available HTTP: <http://www.ams.org/bull/1959-65-04/S0002-9904-1959-10344-X/S0002-9904-1959-10344-X.pdf> (accessed 12 March 2010).

Atiyah, M.F. and Singer, I.M. (1963) 'The index of elliptic operators on compact manifolds', *Bulletin of the American Mathematical Society*, 69: 422–33. Online. Available HTTP: <http://www.ams.org/bull/1963-69-03/S0002-9904-1963-10957-X/S0002-9904-1963-10957-X.pdf> (accessed 10 June 2011).

Atti (1929) *Atti del Congresso internazionale dei matematici, Bologna, 3–10 Settembre 1928*. Vol. 1–6. Bologna: Nicola Zanichelli.

Aubin, D. (1997) 'The withering immortality of Nicolas Bourbaki: A cultural connector at the confluence of mathematics, structuralism and the Oulipo in France', *Science in Context*, 10 (2): 297–342. Online. Available HTTP: <http://www.institut.math.jussieu.fr/~daubin/publis/1997.pdf> (accessed 15 September 2011).

Baum, R. (1994) *Burying Mao: Chinese Politics in the Age of Deng Xiaoping*, Princeton, NJ: Princeton University Press.

Beaulieu, L. (1989) *Bourbaki: une histoire du groupe de mathématiciens français et de ses travaux (1934–1944)*, unpublished PhD dissertation, Université de Montréal.

——(1999) 'Bourbaki's Art of Memory', *Osiris*, 14: 219–51. Online. Available HTTP: <www.jstor.org/stable/301970> (accessed 15 September 2011).

Beeson, M. (2004) 'The mechanization of mathematics', in C. Teuscher (ed.) *Alan Turing: Life and Legacy of a Great Thinker*, Berlin and Heidelberg: Springer. Online. Available HTTP: <http://www.michaelbeeson.com/research/papers/turing2.pdf> (accessed 15 May 2008).

Beijing Normal Institute [Beijing shi yuan shuxue xi jiaogai yi zhi dui Yuquan lu chezhan zhandou xiao zu 北京师院数学系教改一支队玉泉路车场战斗小组, Yuquan lu road depot team of the first education reform corps of the Department of Mathematics of Beijing Normal Institute] (1959) 'Guanyu wuzi diaoyun gongzuo de biao shang zuoye fa 关于物资调运工作的表上作业法' (About the Operation Table Method used in transportation of goods), *Shuxue tongbao* 数学通报, (May 1959): 9–15.

——[Beijing shifan xueyuan shuxue xi san nianji gong nong bing xueyuan 北京师范学院数学系三年级工农兵学员, Third year worker-peasant-soldier students of the Department of Mathematics of Beijing Normal Institute] (1975) 'Ru fa douzheng yu wo guo gudai shuxue de fazhan 儒法斗争与我国古代数学的发展' (The struggle between Confucians and Legalists and the development of our ancient mathematics), *Shuxue xuebao* 数学学报 *(Acta Mathematica Sinica)*, 18 (2): 81–5.

Beijing Normal University [Beijing shifan daxue shuxue xi lilun xuexi xiao zu 北京师范大学数学系理论学习小组, Theoretical study group of the Department of Mathematics of Beijing Normal University] (1975) '*Jiu zhang suan shu* he fa jia luxian 《九章算术》和法家路线' (*Nine Chapters of Mathematical Techniques* and the Legalist line), *Shuxue de shijian yu renshi* 数学的实践与认识 *(Mathematics in Practice and Theory)*, 5 (4): 1–6.

Black, P.E. (1999) 'Chinese postman problem', in V. Pieterse and P.E. Black (eds) *Algorithms and Theory of Computation Handbook*, CRC Press LLC. Online. Available HTTP: <http://xlinux.nist.gov/dads/HTML/chinesePostman.html> (accessed 11 December 2013).

Bledsoe, W.W. and Loveland, D.W. (eds) (1984). *Automated Theorem Proving: After 25 Years*. American Mathematical Society Bookstore.

Blumenthal, O. (1935) 'Lebensgeschichte' (Biography), *David Hilbert – Gesammelte Abhandlungen*, Vol. 3, Berlin: Springer.

Cao Cong (1999) 'The changing dynamic between Science and Politics: Evolution of the highest academic honor in China, 1949–1998', *Isis*, 90 (2): 298–324. Online. Available HTTP: <http://www.jstor.org/stable/237052> (accessed 19 March 2009).

Cartan, H. (1950) 'Une théorie axiomatique des carrés de Steenrod', *C. R. Acad. Sci. Paris*, 230: 425–7.

CAS May 7 School [Zhongguo kexueyuan Hubei Wu Qi xuexiao 中国科学院湖北"五·七"学校] (1971). *Gongzi zhuanyi qingdan* 工资转移清单 (Invoice of forwarded salaries), 13 December 1971, CAS Archives, file Z370-134/12.

Ch'en Po-ta [Chen Boda 陈伯达] (1952) *Speech Before the Study Group of Research Members of Academia Sinica*, Beijing: Foreign Languages Press.

Chang Hao (1987) *Chinese Intellectuals in Crisis: Search for Order and Meaning (1890–1911)*, Berkeley: University of California Press.

Cheatham, T.E., Jr., *et al.* (1973) 'Computing in China. A Travel Report. Computer technology advances rapidly in China with no external aid', *Science*, 182 (4108): 134–40.

Chemla, K. (1987) 'Should they read FORTRAN as if it were English?', *Bulletin of Chinese Studies*, 1 (2): 301–16.

Chemla, K. and Guo Shuchun (2004) *Les Neuf Chapitres*, Paris: Dunod.

Chen Boda 陈伯达 (1971) 'Zai Mao Zedong qizhi xia 在毛泽东旗帜下' (Under Mao Zedong's banner), *Chen Boda wen ji* 陈伯达文集 *(Chen Boda's Collected Writings)*, Vol. 1 (1949–1967), Hong Kong: Lishi ziliao chubanshe.

Chen Cheng-Yih 程贞一 [Joseph Chen] (2010) 'Wu Wenjun yuanshi dui wo yanjiu shuxue shi de qifa he yingxiang 吴文俊原始对我研究数学史的启发和影响' (The inspiration and influence of Academician Wu Wen-Tsun on my study of the history of mathematics), in Jiang Boju 姜伯驹, *et al.* (eds) *Wu Wenjun yu Zhongguo shuxue* 吴文俊与中国数学 *(Wu Wen-Tsun and Chinese Mathematics)*, Singapore: Global Publishing.

Chen Jiangong 陈建功, Cheng Minde 程民德 and Wu Wen-Tsun 吴文俊 (1956) 'Luomaniya de di si ci shuxue dahui he Luomaniya de shuxue 罗马尼亚的第四次数学大会和罗马尼亚的数学' (Fourth Romanian Mathematical Congress and Romanian mathematics), *Shuxue tongbao* 数学通报, 1956 (November): 69–76.

Chen Jianxin 陈建新, Zhao Yulin 赵玉林 and Guan Qian 关前 (1994) *Dangdai Zhongguo kexue jishu fazhan shi* 当代中国科学技术发展史 *(Development of Science and Technology in Modern China)*, Wuhan: Hubei jiaoyu chubanshe.

Chen Jingrun 陈景润 (1973) 'On the representation of a large even integer as a sum of a prime and the product of at most two primes', *Zhongguo kexue* 中国科学 *(Scientia Sinica)*, 16 (2): 157–76.

Chen T. Hsi-en (1961) 'Science, scientists and politics', in Gould, S. H. (ed.) *Sciences in Communist China*, Washington, DC: American Association for the Advancement of Science.

Cheng Minde 程民德 (ed.) (1994) *Zhongguo xiandai shuxuejia zhuan*中国现代数学家传 *(Biographies of Modern Chinese Mathematicians)*. Vol. 1. Nanjing: Jiangsu jiaoyu chubanshe.

——(ed.) (1995) *Zhongguo xiandai shuxuejia zhuan*中国现代数学家传 *(Biographies of Modern Chinese Mathematicians)*. Vol. 2. Nanjing: Jiangsu jiaoyu chubanshe.

——(ed.) (1998) *Zhongguo xiandai shuxuejia zhuan*中国现代数学家传 *(Biographies of Modern Chinese Mathematicians)*. Vol. 3. Nanjing: Jiangsu jiaoyu chubanshe.

——(ed.) (2000) *Zhongguo xiandai shuxuejia zhuan*中国现代数学家传 *(Biographies of Modern Chinese Mathematicians)*. Vol. 4. Nanjing: Jiangsu jiaoyu chubanshe.

——(ed.) (2002) *Zhongguo xiandai shuxuejia zhuan* 中国现代数学家传 *(Biographies of Modern Chinese Mathematicians)*. Vol. 5. Nanjing: Jiangsu jiaoyu chubanshe.

Chengqing 澄清 ['Clarification'] (1958) 'Dui "Jihexue yanjiu duixiang" yi wen de yijian 对"几何学研究对象"一文的意见' (Comments on the article 'Objects of geometrical study'), *Shuxue tongbao* 数学通报, (December 1958): 20–21.

Chern Shiing-Shen [Chen Xingshen 陈省身] (1946) 'Characteristic classes of Hermitian manifolds', *The Annals of Mathematics*, 47 (1): 85–121. Online. Available HTTP: <http://www.jstor.org/stable/1969037> (accessed 24 January 2011).

——(2001) 'On the 2002 Congress', *Notices of the AMS*, 48 (8): 816.

——陈省身 (1941) 'Zhongguo suanxue zhi guoqu yu xianzai 中國算學之過去與現在', *Kexue (Science)* 科學, 25 (5–6): 241–5.

——(1988) 'Zhongyang Yanjiuyuan san nian 中央研究院三年' (Three years in Academia Sinica), *Zhongguo keji shiliao* 中国科技史料, 9 (4): 14–16.

Chinese Academy of Sciences (1959) *Zhongguo kexueyuan 1959 nian kexue jishu yanjiu jihua gangyao xiangmu shuomingshu* 中国科学院1959年科学技术研究计划纲要项目说明书 (Explanation of the projects in the outline 1959 science and technology research plan for CAS), 17 November 1959, CAS Archives, file Z370-51/03.

——(1961) *Zhongguo kexueyuan ganbu renmian* 中国科学院干部任免 (Appointments and dismissals of cadres in CAS), 10 February 1961, CAS Archives, file Z370-74/01.

——(1966). *Yuan gan ren zi 58/66* (66)院干任字第58号: *Mianqu Zheng Zhifu Shuxue suo fusuozhang de zhiwu* 免去郑之辅同志数学所副所长的职务 (Removing comrade Zheng Zhifu from the office of Deputy Director of IMCAS), 6 June 1966, CAS Archives, file Z370-110/17.

——[Shoudu gongren jiefangjun zhu Zhongguo kexueyuan Mao Zedong sixiang xuanchuan dui 首都工人解放军驻中国科学院毛泽东思想宣传队, Mao Zedong Thought Propaganda Troup of the Capital workers and PLA stationed in CAS] (1969). *Yuan xuan ge zi 141/69* (69)院宣革字第141号: *Diao Zhao Weishan tongzhi dao shuxue suo gongzuo shi* 调赵蔚山同志到数学所工作事 (Appointment of Zhao Weishan to IMCAS), 31 October 1969, CAS Archives, file Z370-125/01.

——(1970). *Ke zi 79/70* (70)科字第79号: *Guanyu tongji kexueyuan zai wenhua da geming yilai zhuyao keji chengguo de tongzhi* 关于统计科学院在文化大革命以来主要科技成果的通知 (Circular about compiling statistics of major science and technology results since the start of the Great Proletarian Cultural Revolution), 24 August 1970, CAS Archives, file Z370-133/01.

——(1979). *Ke fa ji zi 1017/79* (79)科发计字1017号: *Guanyu jianli Yingyong shuxue yanjiusuo he Xitong kexue yanjiusuo de qingshi baogao* 关于建立应用数学所和系统科学研究所的请示报告 (Report and permission request for the establishment of the Institute of Applied Mathematics and Institute of Systems Science), 1979, CAS Archives, file Z370-198/02.

——[Zhong gong Zhongkeyuan dangzu 中共中科院党组, CAS CPC Organization] (1981). *Ke fa dang zi 043/81* (81)科发党字第043号: *Dui Zou Xiecheng tongzhi wenti de shencha jielun* 对邹协成同志问题的审查结论 (Conclusions of investigations into the problem of comrade Zou Xiecheng), 23 April 1981, CAS Archives, file Z370-238/04.

Chou Min-Chih (1978) 'The debate on science and the philosophy of life in 1923', *Zhongyang yanjiuyuan Jindai shi yanjiuo suo jikan* 中央研究院近代史研究所季刊, 7: 557–82. Online. Available HTTP: <http://www.mh.sinica.edu.tw/MHDocument/PublicationDetail/PublicationDetail_1139.pdf> (accessed 12 April 2013).

Chou Shang-Ching [Zhou Xianqing 周咸青] (1988) *Mechanical Geometry Theorem Proving*, Dordrecht: D. Reidel.

——(2010) 'Wu Wenjun xiansheng he jihe dingli zhengming 吴文俊先生与几何定理证明' (Wu Wen-Tsun and geometric theorem proving), in Jiang Boju 姜伯驹, *et al.* (eds) *Wu Wen-Tsun yu Zhongguo shuxue* 吴文俊与中国数学 (*Wu Wen-Tsun and Chinese Mathematics*), Singapore: Global Publishing.

Chow Wei-Liang [Zhou Weiliang 周炜良] (1956) 'Algebraic varieties with rational dissections', *Proceedings of the National Academy of Sciences*, 42: 116–19.

Corry, L. (2004) *Modern Algebra and the Rise of Mathematical Structures*, 2nd edn, Basel: Birkhäuser.

CPC Central Committee [Zhong gong zhongyang weiyuanhui 中共中央委员会] [1961] (1997) 'Zhong gong zhongyang tongyi Nie Rongzhen "Guanyu dangqian ziran kexue gongzuo zhong ruogan zhengce wenti de qingshi baogao" he Guojia kewei dangzu, Zhongguo kexueyuan dangzu "Guanyu ziran kexue yanjiu jigou dangqian gongzuo de shi si tiao yijian (cao'an)" de baogao 中共中央同意聂荣臻《关于当前自然科学工作中若干政策问题的请示报告》和国家科委党组、中国科学院党组《关于自然科学研究机构当前工作的十四条意见（草案）》的报告' (CPC Central Committee approves Nie Rongzhen's 'Report on certain policy problems in present work in the natural sciences' and the 'Fourteen Articles on the present work of natural science research institutions (draft)' by the National Science and Technology Committee Party Organization and the Chinese Academy of Sciences Party Organization), in *Jian guo yi lai zhongyao wenxian xuanbian* 建国以来重要文献选编 *(Selection of Important Documents since the Foundation of the Republic)*, Vol. 14, Beijing: Zhongyang wenxian chubanshe. Online. Available HTTP: <http://cpc.people.com.cn/GB/64184/64186/66668/4493472.html>

CSCPRC [Committee on Scholarly Communication with the People's Republic of China] (1974), *China Exchange Newsletter*, 2 (1): 2.

Cullen, C. (1996) *Astronomy and Mathematics in Ancient China: The Zhou bi suan jing*, Cambridge: Cambridge University Press.

——(2004) *The Suan shu shu 'Writings on reckoning'*. Online. Available HTTP: <http://www.nri.org.uk/SuanshushuC.Cullen2004.pdf> (accessed 22 January 2008).

Daruvala, S. (2000) *Zhou Zuoren and an Alternative Chinese Response to Modernity*, Cambridge, MA: Harvard University Asia Center.

Dauben, J. W. (2007) 'Chinese mathematics', in V. Katz (ed.) *The Mathematics of Egypt, Mesopotamia, China, India, and Islam. A Sourcebook*, Princeton, NJ: Princeton University Press.

Dieudonné, J. (1985) *History of Algebraic Geometry*, Monterey: Wadsworth.

——(1989) *A History of Algebraic and Differential Topology, 1900–1960*, Basel: Birkhäuser.

Dikötter, F. (2010) *Mao's Great Famine: The History of China's most Devastating Catastrophe, 1958–1962*, New York: Walker & Co.

Ding Shisun 丁石孙, Yuan Xiandong 袁向东 and Zhang Zugui 张祖贵 (1994) 'Ji du cangsang liang bin ban, tao li tianxia wei xin tian – Duan Xuefu jiaoshou fangtan lu 几度沧桑两鬓斑,桃李天下慰心田 —— 段学复教授访谈录' (Both temples grey after several vicissitudes, I am jubilant about the blossoming of the world – Interview with Professor Duan Xuefu), *Shuxue de shijian yu renshi* 数学的实践与认识 *(Mathematics in Practice and Theory)*, 23 (4).

Dittmer, L. (1990) 'Patterns of elite strife and succession in Chinese politics', *China Quarterly*, (123): 405–30. Online. Available HTTP: <http://www.jstor.org/stable/654150> (accessed 12 November 2009).

——(1991) 'Learning from trauma: The Cultural Revolution in post-Mao politics', in Joseph, W.A., Wong, C.P.W. and Zweig, D. (eds) *New Perspectives on the Cultural Revolution*, Cambridge, MA: The Council on East Asian Studies.

Dold, A. (1956) 'Vollständigkeit der Wuschen Relationen zwischen den Stiefel-Whitneyschen Zahlen differenzierbarer Mannigfaltigkeiten', *Mathematische Zeitschrift*, 65 (1): 200–206. Online. Available HTTP: <http://dx.doi.org/10.1007/BF01473879> (accessed 19 October 2010).

Du Junfu 杜钧福 (2008) 'Wo suo canjia de wenhua da geming' 我所参加的文化大革命 (The Cultural Revolution on which I participated). Online. Available HTTP: <http://blog.sina.com.cn/s/articlelist_1350072355_3_1.html> (accessed 9 October 2009).

——(2009a) 'Chen Boda daotai zhiqian zhishi gao "lian gang zidonghua" 陈伯达倒台之前指示搞"炼钢自动化"' (On Chen Boda's order to 'automate steel production' prior to his demise), *Jiyi* 记忆 *(Remembrance)*, (27): 27–31. Online. Available HTTP: <http://www.tsinghua.org.cn/alumni/downloadTheolFile.do?id=DBCPDBDCDADADADADADCDHDGDCDBDFCPLMMHNCOELFNKDCDHMGNKCOGEGPGD> (accessed 19 October 2009).

——(2009b) 'Wo shi zenyang bei dacheng "wu yi liu" fenzi de 我是怎样被打成"五一六"分子的' (How I was labelled a 'May 16 Element'), *Jiyi* 记忆 *(Remembrance)*, 30. Online. Available HTTP: <http://www.tsinghua.org.cn/alumni/downloadTheolFile.do?id=DBCPDBDCDADADADADADCDIDEDEDHDBCPLMMHNCOELFNKDDDAMGNKCOGEGPGD> (accessed 19 October 2009).

Du Shiran 杜石然 and Mei Rongzhao 梅榮照 (1982) 'Ping Li Yuese zhu *Zhongguo kexue jishu shi* yi shu de shuxue bufen 评李约瑟著《中国科学技术史》一书的数学部分' (Review of the Mathematics section in *Science and Civilisation in China* by Joseph Needham), *Keji shi wen ji* 科技史文集, 8: 1–9.

Du Songzhu 杜松竹 and Zhang Suochun 张锁春 (2000) *Zhongguo xiandai shuxuejia Qin Yuanxun* 中国现代数学家秦元勋 *(The Modern Chinese Mathematician Qin Yuanxun)*, Guiyang: Guizhou keji chubanshe.

Duara, P. (1995) *Rescuing History from the Nation: Questioning Narratives of Modern China*, Chicago: University of Chicago Press.

Engels, F. [1877–78] (1987) *Anti-Dühring. Herr Eugen Dühring's Revolution in Science*, in *Marx and Engels Collected Works*. Vol. 25. English translation of the 1894 edition, London: Lawrence and Wishart. Online. Available HTTP: <http://www.marxists.org/archive/marx/works/cw/volume25/index.htm> (accessed 2 December 2013).

Esherick, J.W., Pickowicz, P.G. and Walder, A.G. (eds) (2006) *The Chinese Cultural Revolution as History*, Stanford: Stanford University Press.

Eves, H.W. (1972) *Mathematical Circles Squared*, Boston: Prindle, Weber & Schmidt.

Fan Dainian 范岱年 (1997) 'Guanyu Zhongguo jindai kexue luohou yuanyin de taolun 关于中国近代科学落后原因的讨论' (Discussion of the causes for the backwardness of Chinese science in the modern period), *Er shi yi shiji* 二十一世纪 *(Twenty-first Century)*, (44); reprinted in Liu Dun 刘钝 and Wang Yangzong 王杨宗 (eds) (2002) *Zhongguo kexue yu kexue geming* 中国科学与科学革命 *(Chinese Science and the Scientific Revolution)*, Dalian: Liaoning jiaoyu chubanshe 625–43.

Fan Fengqi 范凤岐 (1964) *Ziwo jiancha* 自我检查 (My self-examination), January 1964, CAS Archives, file Z370-92/4.

——范凤岐 and IMCAS (1964) *Ganbu jianding biao* 干部睼定表 (Cadre evaluation form), 20 March 1964, CAS Archives, file Z370-99/3.

Fan Hongye 樊洪业 (ed.) (2000). *Zhongguo kexueyuan biannian shi, 1949–1999* 中国科学院编年史1949–1999 *(Chronicle of the Chinese Academy of Sciences, 1949–1999)*, Shanghai: Shanghai keji jiaoyu chubanshe.

Feuerwerker, A. (1961) 'China's history in Marxian dress', *The American Historical Review*, 66 (2): 323–53. Online. Available HTTP: <http://www.jstor.org/stable/1844030> (accessed 15 May 2011).

Fisher, G.J. and Wing, O. (1966) 'Computer recognition and extraction of planar graphs from the incidence matrix', *IEEE Transactions on Circuit Theory*, 13 (2): 154–63. Online. Available HTTP: <http://ieeexplore.ieee.org/iel5/8147/23390/01082574.pdf?arnumber=1082574> (accessed 14 June 2011).

Fitzgerald, A. and MacLane, S. (eds) (1977). *Pure and Applied Mathematics in the People's Republic of China: A trip report of the American Pure and Applied Mathematics Delegation, submitted to the Committee on Scholarly Communication with the People's Republic of China*, Washington: National Academy of Sciences.

Fu Hailun 傅海伦 (2003) *Chuantong wenhua yu shuxue jixiehua* 传统文化与数学机械化 *(Traditional Culture and Mechanization of Mathematics)*, Beijing: Kexue chubanshe.

Fung, E.S.K. (2010) *The Intellectual Foundations of Chinese Modernity: Cultural and Political Thought in the Republican era*, New York: Cambridge University Press.

Furth, C. (ed.) (1976) *The Limits of Change: Essays on Conservative Alternatives in Republican China*, Cambridge, MA: Harvard University Press.

Gan Danyan 干丹岩 (2005) *Daishu tuopu he weifen tuopu jianshi* 代数拓扑和微分拓扑简史 *(A Short History of Algebraic and Differential Topology)*, Changsha: Hunan jiaoyu chubanshe.

——(2010) 'Huiyi he ganhuai 回忆和感怀' (Memories and gratefulness), in Jiang Boju 姜伯驹, *et al.* (eds) *Wu Wen-Tsun yu Zhongguo shuxue* 吴文俊与中国数学 *(Wu Wen-Tsun and Chinese Mathematics)*, Singapore: Global Publishing.

Gellner, E. (1964) *Thought and Change*, London: Weidenfeld and Nicholson.

——(1983) *Nations and Nationalism*, Oxford: Blackwell.

——(1994) *Encounters with Nationalism*, Oxford and Cambridge, MA: Blackwell.

Geraghty, M.A. and Lin Bor-Luh (1984) 'Topological minimax theorems', *Proceedings of the American Mathematical Society*, 91 (3): 377–80. Online. Available HTTP: <http://www.jstor.org/stable/2045306> (accessed 20 January 2011).

Gingrich, C.H. (1929) 'The International Congress of Mathematics at Bologna', *Popular Astronomy*, 36 (9): 529–32.

Glicksberg, I.L. (1952) 'A further generalization of the Kakutani fixed point theorem, with application to Nash equilibrium points', *Proceedings of the American Mathematical Society*, 3 (1): 170–4. Online. Available HTTP: <http://www.jstor.org/stable/2032478> (accessed 20 January 2011).

Goldman, M. (1987a) 'The Party and the intellectuals', in R. MacFarquhar and J.K. Fairbank (eds) *Cambridge History of China. Vol. 14: The People's Republic, Part 1: The Emergence of Revolutionary China 1949–1965*, Cambridge: Cambridge University Press.

——(1987b) 'The Party and the intellectuals: Phase two', in R. MacFarquhar and J.K. Fairbank (eds) *Cambridge History of China. Vol. 14: The People's Republic, Part 1: The Emergence of Revolutionary China 1949–1965*, Cambridge: Cambridge University Press.

Graham, L.R. and Kantor, J.-M. (2009) *Naming Infinity: A True Story of Religious Mysticism and Mathematical Creativity*, Cambridge, MA: Belknap Press of Harvard University Press.

Gray, J.J. (2007) *Worlds out of Nothing: A Course in the History of Geometry in the 19th Century*, New York: Springer.

Gu Jinyong 顾今用 [Wu Wen-Tsun 吴文俊] (1975) 'Zhongguo gudai shuxue dui shijie wenha de weida gongxian 中国古代数学对世界文化的伟大贡献' (Great contribution of ancient Chinese mathematics to world culture), *Shuxue xuebao* 数学学报 *(Acta Mathematica Sinica)*, 18 (1): 18–23.

Gu Mainan 顾迈南 (2002) 'Huiyi dui Hua Luogeng he Chen Jingrun de caifang 回忆对华罗庚和陈景润的采访' (Reminiscenses on the interviews with Hua Loo-Keng and Chen Jingrun), *Xinwen aihaozhe* 新闻爱好者, 2002 (6): 9–11.

Guan Meigu 管梅谷 (1960) 'Qi-ou dian tu shang zuoye fa 奇偶点图上作业法' (Operation Diagram Method using odd and even points), *Shuxue xuebao* 数学学报 *(Acta Mathematica Sinica)*, 11 (3): 263–6.

Guan Zhaozhi 关肇直 (1955) 'Lun muqian dui shuxue zhe men kexue de ji zhong cuowu kanfa 论目前对数学这门科学的几种镉误看法' (About certain current mistaken views on the science of mathematics), *Kexue tongbao* 科学通报, 1955 (10): 36–40.

——(1956a) 'Canjia di san jie quan Sulian shuxue da hui de yi xie jianwen 参加第三届全苏联数学大会的一些见闻' (Impressions from the Third All-Soviet Mathematical Congress), *Kexue tongbao* 科学通报, 1956 (9): 86–9.

——(1956b) 'Dui yi bu kexue de shuxue shi de yaoqiu 对一部科学的数学史的要求评 D.J.Struik著, A Concise History of Mathematics' (Requirements on a scientific history of mathematics – review of D.J.Struik's *A Concise History of Mathematics*), *Shuxue jinzhan* 数学进展 *(Advances in Mathematics)*, 2 (2): 296–304.

——(1956c) 'Ji di san jie quan Sulian shuxue dahui 及第三届全苏联数学大会' (Notes from the Third All-Soviet Mathematical Congress), *Shuxue jinzhan* 数学进展*(Advances in Mathematics)*, 2 (4): 721–8.

——(1956d). *Guanyu peiyang xueshu kongqi, jianli xueshu chuantong, baozheng zhenzheng zai shi er nian nei zai kexue yanjiu fangmian ganshang guoji xianjin shuiping de yijian* (Suggestions about cultivating scholarly climate, establishing scholarly traditions, and making sure that we can indeed reach advanced world level in scientific research in twelve years), 30 July 1956, CAS Archives, file Z370-29/02.

——(1957) 'Lun shuxue de duixiang 论数学的对象' (About the objects of mathematics), *Ziran bianzhengfa yanjiu tongxun* 自然辩证法研究通讯 *(Research Notices about the Dialectics of Nature)*, 1 (2): 1–8.

——关肇直 (1958a) 'Shi jiu shiji yilai de shuxuejia zai renshilun shang suo biaoxian de zifa weiwuzhuyi qingxiang – shuxue sixiang zhaji yi 十九世纪以来的数学家在认识论上所表现的自发唯物主义倾向 --数学思想史札记(一)' (Spontanous materialism in the epistemology of mathematicians since the nineteenth century – notes from studying history of mathematical thought 1), *Ziran bianzhengfa yanjiu tongxun* 自然辩证法研究通讯, 2 (1): 9–18.

——关肇直 (1958b) 'Jihe xue yu jingyan – shuxue sixiang shi zhaji zhi er 几何学与经验——数学思想史扎记之二' (Geometry and experience – notes from studying the history of mathematical thought 2), *Ziran bianzhengfa yanjiu tongxun* 自然辩证法研究通讯, 2 (2): 1–6.

——(1958c) 'Shuxue yanjiu he jiaoxue ye xuyao hou jin bo gu 数学研究和教学也要厚今薄古' (Mathematical research and education also need more of the present and less of the past), *Renmin ribao* 人民日报 *(People's Daily)*, 16 August 1958: 6; reprinted in *Shuxue tongbao*, October 1958: 3–4.

——(1959) 'Cong jindai shuxue shi kan shengchan shiji duiyu shuxue fazhan de zuoyong 从近代数学史看生产实际对于数学发展的作用' (The role of production practice for the development of mathematics seen from the history of modern mathematics), *Shuxue tongbao* 书学通报, 1959 (8): 2–8.

——关肇直 (1964) 'Fanhan fenxi de laiyuan 泛函分析的来源' (The origin of functional analysis), *Kexue tongbao* 科学通报, 1964 (3): 28–33.

——关肇直 and IMCAS (1964). *Ganbu jianding biao* 干部鍵定表 (Cadre evaluation form), 20 March 1964, CAS Archives, file Z370-99/2.

Guo Jinbin 郭金彬 (1987) 'Dui Zhongguo gudai de "jihe daishuhua" de chubu yanjiu 对中国古代的几何代数化的初步研究' (A preliminary study of 'algebraization of geometry' in ancient Chinese mathematics), presented at *International Symposium on the 740th anniversary of Qin Jiushao's 'Mathematical Treatise in*

Nine Sections', Beijing. (unpublished). Quoted from NRI Offprint Collection, Mathematics XIV.

——郭金彬 and Kong Guoping 孔国平 (2004) *Zhongguo chuantong shuxue sixiang shi (History of Traditional Chinese Mathematical Thought)*, Beijing: Kexue chubanshe.

Guo Lei 郭雷 and Wang Yuefei 王賃飞 (2007) 'Sun Keding tongzhi bugao 孙克定同志讣告' (Obituary of comrade Sun Keding). Online. Available HTTP: <http://www.amss.ac.cn/xwdt/zhxw/2007/200908/t20090804_2310274.html> (accessed 10 May 2011).

Guo Moruo 郭沫若 (1957) 'Zhongguo kexueyuan 1956 niandu kexue jiangjin (ziran kexue bufen) tonggao 中国科学院颁发1956年度科学奖金（自然科学部分）通告' (Announcement of the Science Prize of the Chinese Academy of Sciences (natural sciences) for the year 1956), *Renmin ribao* 人民日报 *(People's Daily)*, 25 January 1957: 1.

——(1958) 'Guanyu hou jin bo gu wenti 关于厚今薄古问题', *Renmin Ribao (People's Daily)*, 11 June 1958: 7.

Guo Shuchun 郭書春 (ed.) (2004) *Hui jiao 'Jiu zhang suan shu' (Collation of Editions of 'Nine Chapters on Mathematical Art')*. 2nd, revised edn, Shenyang: Liaoning jiaoyu chubanshe – Taiwan Jiuzhang chubanshe.

Haefliger, A. (1961) 'Differentiable imbeddings', *Bulletin of the American Mathematical Society*, 67 (1): 109–12. Online. Available HTTP: <http://www.jstor.org/stable/1970475> (accessed 24 November 2010).

Halberstam, H. (1986) 'An obituary of Loo-keng Hua', *Mathematical Intelligencer*, 8 (4): 63–5.

Hao Xinhong 郝新鸿 (2009a) 'Cong kexue zhexue shijiao pingjia Wu Wenjun de xueshu chuangxin yu gongxian 从科学哲学视角评价吴文俊的学术创新与贡献' (Evaluating Wu Wen-jun's academic innovation and contribution from the perspective of Philosophy of Science), *Taiyuan shifan xueyuan xuebao / Shehui kexue ban* 太原师范学院学报 社会科学版 *(Journal of Taiyuan Normal University / Social Sciences)*, 8 (2): 6–11.

——(2009b) 'Kuen yu Wu Wenjun: fan Huige shiguan de bijiao yanjiu 库恩与吴文俊:反辉格史观的比较研究' (Kuhn and Wu Wen-Tsun: A comparative study of the anti-Whig view of history), *Nanjing gongye daxue xuebao / Shehui kexue ban* 南京工业大学学报 社会科学版 *(Journal of Nanjing Industrial University / Social Sciences)*, 8 (4): 16–20.

——(2011). A reflection on cultural relativism in the study of the history of science: A case study of Wu Wenjun's independent innovation. Paper presented at 13th International Conference on the History of Science in East Asia, Hefei, China, 23–27 July 2011.

Harzing, A.-W. and van der Wal, R. (2008) 'Google Scholar: The democratization of citation analysis?', *Ethics in Science and Environmental Politics*, 8 (1): 62–71.

He Taiyou 何太由 and Zhang Zhongliang 张忠良 (1993) *Mao Zedong zhanfa* 毛泽东战法 *(Mao Zedong's Art of War)*, 2nd edn, Beijing: Guofang daxue chubanshe.

Hilbert, D. (1899) *Grundlagen der Geometrie*, Göttingen.

Hirzebruch, F. (1956) *Neue topologische Methoden in der algebraischen Geometrie*, Berlin – Heidelberg: Springer.

Hitchin, N.J. (2006) 'Obituary – Shiing-shen Chern 1911–2004', *Bulletin of the London Mathematical Society*, 38: 507–19. Online. Available HTTP: <http://people.maths.ox.ac.uk/hitchin/hitchinlist/chern.pdf> (accessed 22 September 2011).

Hodge, W.V.D. and Pedoe, D. (1947–54) *Methods of Algebraic Geometry*, Cambridge: Cambridge University Press.

Hoe, J. (2008) *Zhu Shijie: The Jade Mirror of the Four Unknowns*, Christchurch: Mingming Bookroom.

Hoffmann, J.E. (1953–57) *Geschichte der Mathematik*, Berlin: G.J.Göschen.

Horiuchi, A. (2010) *Japanese Mathematics in the Edo Period (1600–1868)*, Basel: Springer.

Hsu, E. (2006) 'Reflections on the "discovery" of the antimalarial qinghao', *British Journal of Clinical Pharmacology*, 61 (6): 666–70. Online. Available HTTP: <http://dx.doi.org/10.1111/j.1365-2125.2006.02673.x> (accessed 6 September 2011).

Hu Danian (2005) *China and Albert Einstein: The Reception of the Physicist and his Theory in China 1917–1979*, Cambridge, MA: Harvard University Press.

Hu Fanfu 胡凡夫 (1980). *Bao yuan dang zu bing Fang Yi tongzhi guanyu Wu Wenjun tongzhi qingkuang* 报院党组并方毅同志关于吴文俊同志情况 (A report to the Academy Party Group and Comrade Fang Yi about Comrade Wu Wen-Tsun), 22 February 1980, CAS Archives, file Z373-6/05.

Hu Huakai 胡化凯 (2006) 'Wen Ge qijian de xiangdui lun pipan "文革"期间的相对论批判' (Criticism of theory of relativity during the Cultural Revolution), *Ziran bianzhengfa tongxun* 自然辩证法通讯 *(Bulletin of the Dialectics of Nature)*, 28 (4): 61–70.

Hu, S.T. (1956), 'Wu, Wen-Tsün. Topological invariants of new type of finite polyhedrons', *Mathematical Reviews Online:* MR0072478. Online. Available HTTP: <http://www.ams.org/mathscinet/> (accessed 13 October 2013).

Hu Sen 胡森 (2007) 'Chen Xingshen yu Zhongguo shuxue qiangguo meng 陈省身与中国数学强国梦' (Chern Shiing-Shen and the dream of China as a mathematical great power), in Wu Wen-Tsun 吴文俊 and Ge Molin 葛墨林 (eds) *Chen Xingshen yu Zhongguo shuxue*, Tianjin: Nankai daxue chubanshe.

Hu Weijia 胡维佳 (2005) 'Cong "You jihua de kaizhan kexue jishu gongzuo" dao di yi ge keji guihua 从"有计划地开展科学技术工作"到第一个科技规划' (From 'Expanding science and technology work in a planned manner' to the first plan for science and technology), *Zhongguo keji shi zazhi* 中国科技史杂志 *(The Chinese Journal for the History of Science and Technology)*, 26 (Supplement): 32–44.

Hu Zuoxuan 胡作玄 and Shi He 石赫 (2002) *Wu Wenjun zhi lu (Wu Wen-Tsun's Road)*, Beijing: Kexue chubanshe.

Hua Loo-Keng 华罗庚 (1951) 'Shuxue shi woguo renmin suo shanchang de xueke 数学是我国人民所擅长的学科' (Mathematics is a discipline in which the Chinese people excels), *Renmin ribao* 人民日报 *(People's Daily)*, 10 February 1951: 3; reprinted in Hua Loo-Keng (2006): *Congming zaiyu qinfen, tiancai zaiyu jilei* 聪明在于勤奋，天才在于积累 *(Intelligence is in Diligence, Talent is in Accumulation)*, Beijing: Zhongguo shaonian ertong xinwen chubanshe.

——(1956) *Cong Yang Hui sanjiao tan qi* 从杨辉三角谈起 *(Starting from Yang Hui's Triangle)*, Beijing: Renmin jiaoyu chubanshe.

——(1957) 'Xiwang wo guo kexue xin sheng liliang hen kuai chengzhang 希望我国科学新生力量很快成长' (We hope that the new born scientific forces of our country grow to maturity fast), *Renmin ribao* 人民日报 *(People's Daily)*, 25 January 1957: 7.

——(1962) *Cong Zu Chongzhi de yuanzhou lü tan qi* 从祖冲之的圆周率谈起 *(Starting from the p of Zu Chongzhi)*, Beijing: Renmin jiaoyu chubanshe.

——(1964a) *Cong Liu Hui de geyuan shu tan qi* 从刘徽的割圆术谈起 *(Starting from Liu Hui's Section of the Circle)*, Beijing: Renmin jiaoyu chubanshe.

——(1964b) *Cong Sunzi de shenqi miao suan tan qi* 从孙子的神奇妙算谈起 *(Starting from the Weird and Wonderful Calculations of Sunzi)*, Beijing: Renmin jiaoyu chubanshe.

——华罗庚 (1964c) *Shuxue guinafa* 数学归纳法 *(Mathematical Induction)*, Shanghai: Shanghai jiaoyu chubanshe.

——华罗庚 and Fan Fengqi 范凤岐 (1959). *Guanyu shuxue yanjiusuo zuzhi tizhi de baogao* 关于数学研究所组织体制的报告 (Report on the organization of IMCAS), 17 February 1959, CAS Archives, file Z370-50/02.

——华罗庚, *et al.* (eds) (1960). *Shi nian lai de Zhongguo kexue – shuxue. 1949–1959* 十年来的中国科学 – 数学 *(1949–1959)* *(Chinese Science in the Past Ten Years – Mathematics (1949–1959))*. Beijing: Zhongguo kexueyuan – Kexue chubanshe.

Huang Zubin 黄祖宾 and Wu Wen-Tsun 吴文俊 (2004) 'Zoujin Wu Wen-Tsun yuanshi 走进吴文俊院士' (Getting to know Academician Wu Wen-Tsun), *Guangxi Minzu Xueyuan xuebao / Ziran kexue ban* 广西民族学院学报　自然科学版 *(Journal of Guangxi University for Nationalities / Natural Sciences)*, 10 (4): 2–5.

Hudecek, J. (2012) 'Ancient Chinese mathematics in action: Wu Wen-Tsun's nationalist historicism after the Cultural Revolution', *East Asian Science, Technology and Society*, 6 (1): 41–64. Online. Available HTTP: <http://easts.dukejournals.org/content/6/1/41.abstract> (accessed 31 March 2012).

IHÉS [Institute des Hautes Études Scientifiques] (1975). *Séminaires de Mathématiques*. Poster in the archives of IHÉS.

IMCAS (1952a) *Gongzi biandong biao* 工资变动表 (Table of salary changes), CAS Archives, file Z370-6/10.

——(1952b) *Shuxue suo chengli hou fazhan fangxiang de yijian yu niandu baogao* 数学所成立后发展方向的意见与年度工作报告 (Suggestions for direction of IMCAS development and [1952] annual report), CAS Archives, file Z370-8/01.

——(1953) *Zhongguo kexueyuan shuxue yanjiusuo gongzuo renyuan huamingce* 中国科学院数学研究所工作人员花名册 (Employee profiles of IMCAS), CAS Archives, file Z370-11/02.

——(1955) *Zhongguo kexueyuan 1954 nian gelei ganbu tongji nianbaobiao* 中国科学院1954年各类干部年报表 (Statistics of all types of cadres for 1954), 1 January 1955, CAS Archives, file Z370-14/02.

——(1957) *Zhongguo kexueyuan 1956 nian gelei ganbu tongji nianbaobiao* 中国科学院1956年各类干部年报表 (Statistics of all types of cadres for 1956), 7 January 1957, CAS Archives, file Z370-27/01.

——(1958a) 'Ji Zhongguo kexueyuan Shuxue yanjiusuo pipan zichan jieji xueshu sixiang 记中国科学院数学研究所批判资产阶级学术思想' (Recording the criticism of bourgeois academic thought in IMCAS), *Shuxue tongbao* 数学通报, (October 1958): 4–6.

——(1958b) 'Zouchu shuzhai, mianxiang shengchan, rang shuxue zai woguo kaichu canlan zhi hua 走出书斋，面向生产，让数学在我国开出灿烂之花' (Walk out from the library, turn to production and let mathematics bloom brilliantly in our country), *Shuxue tongbao* 数学通报, (November 1958): 7–8.

——(1958c) *Muqian quan suo renyuan mingdan biao* 目前全所人员名单表 (List of current staff of the entire institute), 27 November 1958, CAS Archives, file Z370-40/02.

——(1958d) *Shuxue yanjiusuo renyuan zuzhi jigou biao* 数学研究所人员组织机构表 (Organisation of IMCAS staff), 2 December 1958, CAS Archives, file Z370-40/01.

——(1959a) *Shuxue yanjiusuo 1959 nian gongzuo jihua* 数学研究所1959年工作计划 (IMCAS work plan for 1959), 21 February 1959, CAS Archives, file Z370-50/01.

——(1959b) *Zhongguo Kexueyuan 1958 nian ge lei ganbu dingqi tongji baobiao* 中国科学院1958年各类干部定期统计表 (Statistics of all types of CAS cadres for 1958), 28 February 1959, CAS Archives, file Z370-43/1.

——(1959c) Z370-49/07: *Shuxue yanjiusuo gongzuo renyuan shengzhi he shengji shenpi biao* 数学研究所工作人员升职和升定级审批表 (Application forms for promotion and change of rank of IMCAS employees), November 1959, CAS Archives, file.

——(1960a) *1959 nian gongzuo zongjie he 1960 nian de zhuyao renwu (cao'an)* 1959年工作总结和1960年主要任务（草案） (Summary of 1959 and major tasks for 1960 (draft)), 1960, CAS Archives, file Z370-50/03.

——(1960b) *Zhongguo Kexueyuan 1960 nian ge lei ganbu dingqi tongji baobiao* 中国科学院1960年各类干部定期统计表 (Statistics of all types CAS cadres for 1960), 28 December 1960, CAS Archives, file Z370-55/05.

——(1961) *1961.3.29 huibao bianzhi hou qingkuang* 1961.3.29汇报编制后情况 (Report on the situation after quota assignment on 29 March 1961), 29 March 1961, CAS Archives, file Z370-68/04.

——(1962a) *Quan suo renyuan zhu nian jibie biandong qingkuang biao* 全所人员逐年级别变动情况表 (Table of rank changes in the entire institute by years), CAS Archives, file Z370-77/04.

——(1962b) *Fenpei gei ge sheng (qu) de ganbu dengji biao* 分配给各省（区）的干部登记表 (Registration forms for cadres to be assigned to other provinces and regions), CAS Archives, file Z370-78/01.

——(1963) *Jingjian xiafang de renyuan mingdan* 精简下放的人员名单 (List of people sent down after downsizing), September 1963, CAS Archives, file Z370-78/10.

——(1964) *1964 nian quan guo ganbu dingqi tongji biao* 1964年全国干部定期统计表 (National statistics of cadres for 1964), 7 November 1964, CAS Archives, file Z370-96/06.

——(1965) *1965 nian yanjiu jihua timu jianbiao* 一九六五年研究计划题目简表 (Planned research topics for 1965), 23 April 1965, CAS Archives, file Z370-108/1.

——(1966a) *Zhigong renshu zengjian biandong qingkuang tongji jibao ji ge lei renyuan jidu tongji baobiao* 职工人数增减变动情况统计季报及各类人员季度统计报表 (Quarterly statistics of changes in employee numbers and quarterly overall personnel statistics), CAS Archives, file Z370-110/3.

——(1966b) *Kelaya, Naqiao xuexi qingkuang ji jihua* 克拉雅、那翘学习情况及计划 (Study report and plan for Kelaya and Naqiao), September 1966, CAS Archives, file Z370-114.

——(1966c) *Yanjiusheng fenpei dengji biao* 研究生分配登记表 (Registration forms for the assignment of graduate students), CAS Archives, file Z370-114/06.

——(1966d) *Shuxuesuo 1966 nian yanjiu jihua jianbiao* 数学所一九六六年研究计划简表 (Research plan of IMCAS for 1966), CAS Archives, file Z370-115/1.

——(1969) *Gongzi tongzhidan* 工资通知单 (Payrolls), CAS Archives, file Z370-123/01.

——(1970a) *Shuxue suo huaming ce* 数学所花名册 (Personnel list of IMCAS), 1 May 1970, CAS Archives, file Z370-127/04.

——(1970b) *Di si pi xiafang renyuan gongzi qingdan* 第四批下放人员工资清单 (Invoice for salaries of the fourth group of sent-down staff), 15 July 1970, CAS Archives, file Z370-126/02.

——[IMCAS Party Reconstruction Group 所政党建小组 and IMCAS Revolutionary Committee 所革命委员会] (1972a) *1971 niandu gongzuo jiben zongjie he 1972 nian gongzuo yaodian* 数学所1971年度工作总结及1972年工作要点 (Annual summary of the 1971 work and main points for 1972), 10 March 1972, CAS Archives, file Z370-132/01.

——(1972b) *Wenhua da geming zhong qing dui zheng dang qingkuang dengji biao* 文化大革命中清队整党情况登记表 (Registration forms for cleaning the class ranks and strengthening the Party during the Cultural Revolution), 2 September 1972, CAS Archives, file Z370-183/10.

——(1973a) *1972 niandu gongzuo zongjie* 1972年度工作总结 (Annual summary of the 1972 work), CAS Archives, file Z370-140/01.

——(1973b) *1973 nian shang ban nian keyan, shengchan jihua zhixing qingkuang biao* 一九七三年上半年科研、生产计划执行情况表 (Tables of plan fulfilment for the first half of 1973), 13 August 1973, CAS Archives, file Z370-146/03.

——(1973c) *Guanyu Xu Mingwei 'Wu Yiliu' fan geming zuixing de shencha baogao* 关于胥鸣伟"五·一六"反革命罪行的审查报告 (Investigation report of Xu Mingwei's 'May 16' counterrevolutionary crimes), 19 September 1973, CAS Archives, file Z370-238/02.

——[Zhongguo kexueyuan shuxue suo weifen fancheng yanjiushi da pipan zu 中国科学院数学所微分方程研究室大批判组, Criticism Group of Differential Equations Research Division of IMCAS] (1974) 'Chedi pipan "ke ji fu li" de fandong gangling 彻底批判"克己复礼"的反动纲领' (Thoroughly condemn the reactionary outline of 'suppression of the self and restitution of rites'), *Shuxue xuebao* 数学学报 *(Acta Mathematica Sinica)*, 17 (1): 1–4.

——(1975) *Wu Wenjun tongzhi de zhengzhi biaoxian* 吴文俊同志的政治表现介绍 (Political conduct of Comrade Wu Wen-Tsun), CAS Archives, file Z370-157/04.

——(1978) *Wenhua da geming shou shencha qijian yin bing siwang, feizhengchang siwang deng dengji biao* 文化大革命受审查时因病、非正常死亡等登记表 (Records of those who died from illness or for extraordinary reasons during investigation in the Cultural Revolution), 30 March 1978, CAS Archives, file Z370-183/10.

IMCAS CPC Steering Group 中共数学所领导小组 (1976a) *Fanji youqing fan'an feng jianbao 2* 反击右倾翻案风简报2 (Brief report on the counter-attack against the right-opportunist wind to reverse verdicts, no. 2), 19 February 1976, CAS Archives, file Z370-166/02.

——(1976b) *Yundong jianbao 11* 运动简报11 (Brief report on the movement, no. 11), 19 April 1976, CAS Archives, file Z370-166/02.

——(1976c) *Jianbao 15* 简报15 (Brief report, no. 15), 18 May 1976, CAS Archives, file Z370-166/02.

——(1976d) *Jianbao 22* 简报22 (Brief report, no. 22), 6 September 1976, CAS Archives, file Z370-166/02.

——(1976e) *Jianbao 25* 简报25 (Brief report, no. 25), 23 September 1976, CAS Archives, file Z370-166/02.

——(1976f) *Yundong jianbao 2* 运动简报2 (Brief report on the movement, no. 2), 9 November 1976, CAS Archives, file Z370-166/02.

——(1976g) *Yundong jianbao 10* 运动简报10 (Brief report on the movement, no. 10), 24 December 1976, CAS Archives, file Z370-166/02.

IMCAS Linear Programming Group [Zhongguo kexueyuan Shuxue yanjiusuo xianxing guihua xiaozu 中国科学院数学研究所线性规划小组] (1958a) 'Wuzi diaoyun gongzuo zhong de xianjin fangfa – Tu shang zuoye fa 图上作业法' (An advanced method for transportation of goods – the Operation Diagram Method), *Shuxue tongbao* 数学通报, (November 1958): 2–4.

——[Zhongguo kexueyuan Shuxue yanjiusuo xianxing guihua xiaozu 中国科学院数学研究所线性规划小组] (1958b) 'Wuzi diaoyun gongzuo zhong de biao shang zuoye fa 物资调运工作中的表上作业法' (The Operation Table Method in transportation of goods), *Shuxue tongbao* 数学通报, (December 1958): 3–14.

IMCAS Party Steering Group [Zhongguo kexueyuan Shuxue yanjiusuo dang de lingdao xiao zu 中国科学院数学研究所党的领导小组] (1976) *1976 nian shuxue suo keyan gongzuo jihua yaodian* 一九七六年数学所科研工作计划要点 (Main points of the IMCAS research work plan for 1976), 15 April 1976, CAS Archives, file Z370-170/2.

——(1978) *1977 nian gongzuo huibao tiyao (caogao)* 一九七七年工作汇报提要（草稿） (Outline work report for 1977 (Draft)), 6 January 1978, CAS Archives, file Z370-174/01.

——(1981) *Guanyu Xu Mingwei tongzhi wenti de fucha baogao* 关于胥鸣伟同志问题的复查报告 (Reinvestigation report of the problem of comrade Xu Mingwei), 21 June 1981, CAS Archives, file Z370-238/02.

IMCAS PDE Group, Shuxue yanjiusuo pianwei fangcheng zu 数学研究所偏微方程组 (1958) 'Sulian zhuanjia Bichajie tongzhi zai Zhongguo jiangxue de jingguo 苏联专家比察捷在中国啲学的经过' (The lecturing tour of the Soviet expert comrade Bitsadze in China), *Shuxue jinzhan*数学进展 *(Advances in Mathematics)*, 4 (3): 404–9.

IMCAS Revolutionary Committee [Zhongguo kexueyuan Shuxuesuo geming weiyuanhui 中国科学院数学所革命委员会] (1971) *Wenhua da geming yilai kexue jishu chengguo baobiao* 文化大革命以来科学技术成果报表 (Report of science and technology results since the start of the Cultural Revolution), 18 February 1971, CAS Archives, file Z370-133/02.

IMCAS Steering Group [Zhongguo kexueyuan Shuxue yanjiusuo lingdao xiao zu 中国科学院数学研究所领导小组] (1974) *Guanyu Ding Ruimin zisha shijian de baogao* 关于丁瑞敏自杀事件的报告 (Investigation of the suicide of Ding Ruimin), 2 May 1974, CAS Archives, file Z370-149/01.

——(1975a) *1974 nian gongzuo zongjie ji 1975 nian gongzuo yaodian* 一九七四年工作总结及一九七五年工作要点 (Summary of the 1974 work and main points for 1975), January 1975, CAS Archives, file Z370-151/01.

——(1975b) *Shuxue suo qingkuang huibao* 数学所情况汇报 (Situation report for IMCAS), 25 July 1975, CAS Archives, file Z370-161/01.

——(1976) *1975 nian gongzuo zongjie* 一九七五年工作总结 (Summary of the 1975 work), 1 February 1976, CAS Archives, file Z370-156/01.

Institute of Mathematics of Academia Sinica (Shanghai) (1947) *Zhongyang yanjiuyuan shuxue yanjiusuo renyuan renmian qiandiao kaoji xinji wenshu* 中央研究院数学研究所人员任免迁调考绩薪给文书 (Documents on personnel matters, evaluation and salaries of the Institute of Mathematics, Academia Sinica), Second Historical Archives, file 393/1666.

Jackson, A. and Kotschick, D. (1998) 'Interview with Shiing Shen Chern', *Notices of the AMS*, 45 (7): 860–5. Online. Available HTTP: <http://www.ams.org/notices/199807/chern.pdf> (accessed 22 September 2011).

Jia Hepeng and Li Jiao (2007) 'Faking it: The debate over Chinese pseudoscience', *Science and Development Network*, 21 February 2007. Online. Available HTTP: <http://

www.scidev.net/en/features/faking-it-the-debate-over-chinese-pseudoscience.html>
(accessed 5 September 2011).

Jiang Boju 姜伯驹, *et al.* (eds) (2010). *Wu Wenjun yu Zhongguo shuxue* 吴文俊与中国
数学 *(Wu Wen-Tsun and Chinese Mathematics)*, Singapore: Global Publishing.

Jiang Xiaoyuan 江晓源 and Liu Bing 刘兵 (2007) *Nan qiang bei diao: kexue yu wenhua
zhi guanxi de duihua*, 南腔北调: 科学与文化之关系的对话 *(Southern Accents,
Northern Tones: Dialogues on the Relationship between Science and Culture)*, Beijing:
Beijing daxue chubanshe.

Jiang Yan 姜岩 (2003) 'Ganshou da shi fengcai – Ji Wu Wenjun yuanshi 感受大师
风采 —— 记吴文俊院士' (Experiencing Master's charisma), *Keji xinxi* 科技信息,
2003 (10).

Jiang Zehan 江泽涵, *et al.* (1958) 'Shuxue gongzuozhe mantan sixiang gaizao 数学
工作者漫谈思想改造' (Mathematical workers talking about the thought reform),
Shuxue tongbao 数学通报, (March 1958).

Jiaotong University History Group [Jiaotong daxue xiao shi bianxie zu 交通大学校
史编写组] (1986) *Jiaotong daxue xiao shi* 交通大学校史 *(History of the Jiaotong
University)*, Shanghai: Shanghai jiaoyu chubanshe

Jing Bao 晶报 (2010), 'Ye Shuwu – Dongfang hong yi hao weixing shang tian de
gongchen 叶述武: 东方红一号卫星上天的功臣' (Ye Shuwu – the servant with
great merit for the launch of the satellite East is Red 1). Online. Available HTTP:
<http://bbs.southcn.com/thread-654709-1-1.html> (accessed 15 November 2013).

Jing Zhujun 井竹君 (1981) *Zhi Yuan suo lingdao tongzhi de xin* 至院所领导同志的信
(A letter to leading comrades in the Academy and the Institute), 8 June 1981, CAS
Archives, file Z370-238/05.

Jinhua Teachers College, Department of Mathematics [Jinhua shi zhuan shuxue ke
金华师专数学科] (1972) 'Jihexue de fazhan yu gongli fangfa 几何学的发展与公理
方法' (Development of geometry and the axiomatic method), *Shuxue de shijian yu
renshi* 数学的实践与认识 *(Mathematics in Practice and Theory)*, 2 (4): 46–50.

Johnston, A.I. (1996) 'Cultural realism and strategy in Maoist China', in P.J.
Katzenstein (ed.) *The Culture of National Security: Norms and Identity in World
Politics*, New York: Columbia University Press. Online. Available HTTP: <http://
www.ou.edu/uschina/texts/Johnston1996CulturalRealismMao.pdf> (accessed 21
August 2011).

Joseph, W.A., Wong, C.P.W. and Zweig, D. (1991) *New Perspectives on the Cultural
Revolution*, Cambridge, MA and London: The Council on East Asian Studies.

Ke Linjuan 柯琳娟 (2009) *Wu Wenjun zhuan: Rang shuxue huigui Zhongguo* 吴文
俊传: 让数学回归中国 *(Biography of Wu Wen-Tsun: Let Mathematics Return to
China)*, Nanjing: Jiangsu renmin chubanshe.

Kindler, J. (1993) 'Topological intersection theorems', *Proceedings of the American
Mathematical Society*, 117 (4): 1003–11.

Knight, N. (2007) *Rethinking Mao: Explorations in Mao Zedong's Thought*, New
York: Lexington Books.

Knuth, D.E. (1972) 'Ancient Babylonian algorithms', *Communications of the
Association for Computing Machinery*, 15 (7): 671–7.

Kwok, D.W.Y. (1965) *Scientism in Chinese Thought 1900–1950*, New Haven: Yale
University Press.

Lam Lay Yong (1997) 'Zhang Qiujian Suanjing (the mathematical classic of Zhang
Qiujian): An overview', *Archive for History of Exact Sciences*, 50 (3–4): 201–40.

Lam Lay Yong and Ang Tian Se (2004) *Fleeting Footsteps*, Singapore: World Scientific.

Lardy, N. (1987) 'The Chinese economy under stress, 1958–1965', in R. MacFarquhar and J.K. Fairbank (eds) *Cambridge History of China, Vol. 14, The People's Republic, Part 1: The Emergence of Revolutionary China 1949–1965*, Cambridge: Cambridge University Press.

Lehto, O. (1998) *Mathematics Without Borders: A History of the International Mathematical Union*, New York and Berlin: Springer. Online. Available HTTP: <http://www.mathunion.org/fileadmin/ICM/Lehto/Lehto_Mathematics_Without_Borders/Lehto_Mathematics_Without_Borders.pdf> (accessed 29 December 2013).

Lemaire, J.-M. (1979), 'Wu, Wen Tsün [Wu, Wen Jun] Theory of I*-functor in algebraic topology – real topology of fiber squares', *Mathematical Reviews Online*: MR0645386 Online. Available HTTP: <http://www.ams.org/mathscinet/> (accessed 23 November 2013).

——(1981), 'Wu, Wen Jun: On calculability of I*-measure with respect to complex-union and other related constructions', *Mathematical Reviews Online*: MR0579833. Online. Available HTTP: <http://www.ams.org/mathscinet/> (accessed 23 November 2013).

Lenin, V.I. [1915] (1976) 'On the Question of Dialectics', in *Lenin Collective Works*, Vol. 38, Moscow: Progress Publishers. Online. Available HTTP: <http://www.marxists.org/archive/lenin/works/1915/misc/x02.htm> (accessed 20 November 2013).

Leray, J. (1945) 'Sur la forme des espaces topologiques et sur les points fixes des représentations', *Journal de Mathématiques Pures et Appliquées*, 24: 95–167.

Levenson, J.R. (1968) *Confucian China and its Modern Fate; A Trilogy*, 1st combined edn, Berkeley: University of California Press.

Li Banghe 李邦河 (1989) 'Wu Wen-Tsun – An outstanding mathematician', *Acta Mathematica Scientia*, 9 (4): 361–5.

——(2010) 'Wu Wenjun dui tuopuxue de weida gongxian 吴文俊对拓扑学的伟大贡献' (Wu Wen-Tsun's great contributions to topology), in Jiang Boju 姜伯驹, *et al.* (eds) *Wu Wenjun yu Zhongguo shuxue* 吴文俊与中国数学 (*Wu Wen-Tsun and Chinese Mathematics*), Singapore: Global Publishing.

Li Di 李迪 (1975) 'Chedi pipan guanyu shuxue qiyuan yu "He tu Luo shu" de miulun 彻底批判关于数学起源于"河图洛书"的谬论' (A thorough criticism of the absurd theory that mathematics originated from the diagrams of He and Luo), *Shuxue de shijian yu renshi* 数学的实践与认识 (*Mathematics in Practice and Theory*), 5 (2): 11–15.

——(1984) *Zhongguo shuxue shi jian bian* 中国数学史简编 (*A Short History of Chinese Mathematics*), Shenyang: Liaoning renmin chubanshe.

——(1975) '*Jiu Zhang Suan Shu* de xingcheng yu xian Qin zhi Xi Han shiqi de ru fa douzheng 《九章算术》的形成与先秦至西汉的儒法斗争' (The formation of the *Nine Chapters* and the struggle between the Confucians and the Legalists from pre-Qin to the Western Han), *Shuxue xuebao* 数学学报 (*Acta Mathematica Sinica*), 18 (4): 223–30.

Li Jimin 李继闵 (1984) 'Shi lun Zhongguo chuantong shuxue de tedian 试论中国传统数学的特点' (Preliminary discussion of the characteristics of traditional Chinese mathematics), *Zhongguo shuxue shi lunwenji* 中国数学史论文集, 2 (1986): 9–18.

——(1987) 'Zhongguo gudai bu ding fenxi de chengjiu yu tese 中国古代不定分析的成就与特色' (Achievements and characteristic features of Chinese ancient indeterminate analysis), in Wu Wen-Tsun 吴文俊 (ed.) *Qin Jiushao yu Shu shu jiu zhang (Qin Jiushao and the Mathematical Treatise in Nine Section)* 秦九韶与《数书九章》, Beijing: Beijing shifan daxue chubanshe.

Li Shihui 李世辉 (2005) 'Keji zizhu chuangxin yu Zhong Xi wenhua hubu wo jian–liu ge dianxing shili de sikao 科技自主创新与中西文化互补之我见 – 六个典型实例的思考' (Personal viewpoint on independent innovation in science and technology, and on the complementarity of Chinese and Western cultures – Discussion of six representative cases), *Zhongguo gongcheng xue* 中国工程学 *(Engineering Science)*, 7 (4): 11–21.

Li Wenlin 李文林 (1991) 'Lun gudai yu zhongshiji de Zhongguo suanfa 论古代于中世纪的中国算法' (On Chinese algorithms in ancient and medieval times), *Shuxue shi yanjiu wenji* 数学史研究文集, 2 (1991): 1–5.

——(2001) 'Gu wei jin yong de dianfan – Wu Wenjun jiaoshou de shuxue shi yanjiu 古为今用的典范 – 吴文俊教授的数学史研究' (A paragon of making past serve the present – Professor Wu Wen-Tsun's studies on the history of mathematics), in Lin Dongyue 林东岳, Li Wenlin 李文林 and Yu Yanlin 虞言林 (eds) *Shuxue yu shuxue jixiehua* 数学与数学机械化 *(Mathematics and Mathematics Mechanization)*, Jinan: Shandong jiaoyu chubanshe.

——(2005) *Shuxue de jinhua: Dong-Xifang shuxue shi bijiao yanjiu* 数学的进化：东西方数学史比较研究 *(Evolution of Mathematics: Comparative Studies of Eastern and Western Mathematics)*, Beijing: Kexue Chubanshe.

——(2008) 'Wu Long 吴龙' (The 'Wu Dragon'), in Shi Jihuai 史济怀 (ed.) *Zhongguo kexue jishu daxue shuxue wu shi nian* 中国科学技术大学数学五十年 *(Fifty Years of Mathematics at the University of Science and Technology of China)* Hefei: Zhongguo kexue jishu daxue chubanshe.

——(2010) 'Gu wei jin yong, zizhu chuangxin de dianfan – Wu Wenjun yuanshi de shuxue shi yanjiu 古为今用、自主创新的典范 – 吴文俊院士的数学史研究' (The paragon of using past for the present and of independent innovation – Academician Wu Wen-Tsun's studies of the history of mathematics), in Jiang Boju 姜伯驹, *et al.* (eds) *Wu Wenjun yu Zhongguo shuxue* 吴文俊与中国数学 *(Wu Wen-Tsun and Chinese Mathematics)*, Singapore: Global Publishing.

——李文林 and Yuan Xiandong 袁向东 (1982) 'Zhongguo gudai buding fenxi ruogan wenti tantao 中国古代不定分析若干问题探讨' (On some problems about indeterminate analysis in ancient and medieval China), *Keji shi wen ji* 科技史文集, 8: 106–22.

Li Xiangdong 李向东 and Zhang Tao 张涛 (2006). *Da jia: Wo de budengshi* 大家：我的不等式 *(Great Masters: My Inequalities)*. TV interview, China Central Television – CCTV.

Li Yaming 李雅明 (ed.) (2005) *Guan Weiyan koushu lishi huiyi lu* 管惟炎口述历史回忆录管惟炎口述历史回忆录 *(Oral History and Reminiscences of Guan Weiyan)*, Hsinchu: National Tsing Hua University Press.

Li Yan 李俨 and Du Shiran 杜石然 (1963–64) *Zhongguo gudai shuxue jian shi* 中国古代数学简史 *(A Concise History of Ancient Chinese Mathematics)*, Beijing: Zhonghua shuju.

Li Zhenzhen 李真真 (2002) 'Queding "lilun yu shiji guanxi" de kunjing: zhengzhi zhixiang yu fazhan zhixiang de chongtu 确定"理论与实际关系"的困境 ：政治指向与发展指向的冲突' (Dilemma about defining 'the relationship between theory and practice': The conflict between political and developmental directions), *Ziran kexue shi yanjiu* 自然科学史研究 *(Studies in the History of Natural Sciences)*, 21 (1): 33–48.

Liao Shan-Dao 廖山涛 (1957) 'Periodic transformations and fixed point theorems. I. Cup products and special cohomology', *Acta Scientiarum Naturalium Universitatis Pekinensis*, 3 (1): 1–38.

Liao Yuqun (2006) *Traditional Chinese Medicine*, Beijing: Wuzhou chuanbo chubanshe.

Lieberthal, K. (1987) 'The Great Leap Forward and the split in Yenan leadership', in R. MacFarquhar and J.K. Fairbank (eds) *Cambridge History of China, Vol. 14, The People's Republic, Part 1: The Emergence of Revolutionary China, 1949–1965*, Cambridge: Cambridge University Press.

Liu Dun 刘钝 (1993) *Da zai yan shu*, 大哉言数 (*Discussing Mathematics – What a Vast Subject*), Dalian: Liaoning jiaoyu chubanshe.

——(2000) 'A new survey of the Needham question', *Ziran kexue shi yanjiu* 自然科学史研究 (*Studies in the History of Natural Sciences*), 19 (4): 293–305.

——刘钝 and Wang Yangzong 王扬宗 (eds) (2002) *Zhongguo kexue yu kexue geming* 中国科学与科学革命 (*Chinese Science and the Scientific Revolution*), Dalian: Liaoning jiaoyu chubanshe.

Liu Haifan 刘海藩 and Wan Fuyi 万福义 (2006) *Mao Zedong sixiang zonglun* 毛泽东思想综论 (*Introduction to Mao Zedong Thought*), Beijing: Zhongyang wenxian chubanshe.

Liu, L. He (1995) *Translingual Practice: Literature, National Culture, and Translated Modernity – China, 1900–1937*, Stanford, CA: Stanford University Press.

Liu Qiuhua 刘秋华 (2010) *Er shi shiji zhong wai shuxue sixiang jiaoliu* 二十世纪中外数学思想交流 (*Communication of Mathematical Ideas between China and the World in the Twentieth Century*), Beijing: Kexue chubanshe.

Liu Zhenkun 刘振坤 (1999a) 'Chun feng qiu yu er shi nian – Du Runsheng fangtan lu 春风秋雨二十年–杜润生访谈录' (Twenty years of storms and winds – Interview with Du Runsheng), *Bai nian chao* 百年潮 (*Hundred Year Tide*), 3 (6): 11–22. Online. Available HTTP: <http://www.eywedu.com/Bainianchao/banc1999/banc19990602.html> (accessed 12 June 2009).

——(1999b) 'Zai Kexueyuan huihuang de beihou – Zhang Jingfu fangtan lu 在科学院辉煌的背后–张劲夫访谈录' (Behind the brilliance of the Academy – Interview with Zhang Jingfu), *Bai nian chao* 百年潮 (*Hundred Year Tide*), 3 (6): 1–10. Online. Available HTTP: <http://www.eywedu.com/Bainianchao/banc1999/banc19990601.html> (accessed 10 June 2009).

Lorentz, G.G. (2002) 'Mathematics and Politics in the Soviet Union from 1928 to 1953', *Journal of Approximation Theory*, 116 (2): 169–223. Online. Available HTTP: <http://www.sciencedirect.com/science/article/pii/S0021904502936701> (accessed 26 August 2011).

Lu Jiaxi 卢嘉锡 (ed.) (1992) *Zhongguo xiandai kexuejia zhuanji* 中国现代科学家传记 (*Biographies of Modern Chinese Scientists*). Vol. 3. Beijing: Kexue chubanshe.

Lu Qikeng 陆启铿 (2008) 'Yu jichu yanjiu he xueke jiaocha you guan de yixie jingyan 与基础研究和学科交叉有关的一些经验' (Some experiences with fundamental research and crossing of disciplinary boundaries), *Kexue shibao* 科学时报 (*Science Times*), 25 July 2008: A2. Online. Available HTTP: <http://www.sciencenet.cn/dz/dznews_photo.aspx?id=3938> (accessed 28 October 2009).

——(2010) 'Wu Wenjun xiansheng de xueshu sixiang dui wo de yingxiang 吴文俊先生的学术思想对我的影响' (The influence on me of Wu Wen-Tsun's academic thought), in Jiang Boju 姜伯驹, *et al.* (eds) *Wu Wenjun yu Zhongguo shuxue* 吴文俊与中国数学 (*Wu Wen-Tsun and Chinese Mathematics*), Singapore: World Scientific.

Luo Haipeng 罗海鹏 (2005), 'Qing Guan Zhaozhi zuotan he pipan Hua Luogeng (Huiyi Gao Jianguo tongxue)' 请关肇直座谈与批判华罗庚（回忆高建国同学）(Asking Guan Zhaozhi to give a talk and criticizing Hua Luogeng (Reminiscences of Gao Jianguo)). Online. Available HTTP: <http://alumni.ustc.edu.cn/view_notice.php?msg_id=904> (accessed 23 April 2009).

Luo Shengxiong 罗声雄 (2001) *Yi ge zhenshi de Chen Jingrun* 一个真实的陈景润（*A Real Chen Jingrun*), Beijing: Chang Jiang wenyi chubanshe.

MacFarquhar, R. (1983) *Origins of the Cultural Revolution*, Vol. 2: *The Great Leap Forward, 1958–1960*, New York: Columbia University Press.

——(1997) *Origins of the Cultural Revolution*, Vol. 3: *The Coming of the Cataclysm, 1961–1966*, Oxford: RIIA.

MacFarquhar, R. and Fairbank, J.K. (eds) (1987) *The People's Republic, Part 1: The Emergence of Revolutionary China 1949–1965*. Cambridge History of China. Vol. 14. Cambridge: Cambridge University Press.

MacFarquhar, R. and Schoenhals, M. (2006) *Mao's Last Revolution*, Cambridge, MA and London: Belknap Press of Harvard University Press.

MacLane, S. (1980) 'Pure and applied mathematics', in L.A. Orleans (ed.) *Science in Contemporary China*, Stanford: Stanford University Press.

MacPherson, R.D. (1974) 'Chern classes for singular algebraic varieties', *Annals of Mathematics*, 100 (2): 423–32. Online. Available HTTP: <http://www.jstor.org/stable/1971080> (accessed 12 April 2010).

Mao Tse-Tung [Mao Zedong 毛泽东] (1965) *Selected Works*, Vol. 1–4, Peking: Foreign Languages Press.

——[1965] (1994). You Fight Your Way And I'll Fight My Way – A Conversation With The Palestine Liberation Organization Delegation, in *Selected Works of Mao Tse-Tung*, Vol. 9, Hyderabad: Sramikavarga Prachuranalu. Online. Available HTTP: <http://www.marxists.org/reference/archive/mao/selected-works/volume-9/mswv9_40.htm> (accessed 20 August 2011).

Mao Zedong 毛泽东 [1937] (1960) 'Shijian lun 实践论' (On Practice), in *Mao Zedong xuanji*, Vol. 1, Beijing: Renmin chubanshe.

——[1930] (1960) Xingxing zhi huo, keyi liao yuan 星星之火，可以燎原 (A single spark can start a prairie fire), in *Mao Zedong xuanji*, Vol. 1, Beijing, Renmin chubanshe.

——[1938] (1960) Tongyi zhanxian zhong de duli zizhu wenti 统一战线中的独立自主问题 (The question of independence and initiative within the United Front), in *Mao Zedong xuanji*, Vol. 2, Beijing: Renmin chubanshe.

——(1993–9) *Mao Zedong wen ji,* 毛泽东文集（*Mao Zedong's Collected Works*), Vol. 1–8, Beijing: Renmin chubanshe.

——[1947] (1996). Xian da ruo de hou da qiang de, ni da ni de, wo da wo de 先打弱的后打强的你打你的我打我的 (Fight the weak at first and the strong later, you fight your way and I fight my way), in *Mao Zedong wenji*, Vol. 4, Beijing: Zhongyang wenxian chubanshe.

——[1950] (1999). Chaoxian zhanju he women de fangzhen 朝鲜战局和我们的方针 (The war situation in Korea and our pointers), in *Mao Zedong wenji*, Vol. 6.

——(2010) *Jianguo yilai Mao Zedong junshi wengao,* 建国以来毛泽东军事文稿（*Mao Zedong's Post-Liberation Military Essays*), Beijing: Junshi kexue chubanshe – Zhongyang wenxian chubanshe.

Martzloff, J.-C. (1997) *A History of Chinese Mathematics*, Berlin Heidelberg: Springer.

Mathias, A.R.D. (1992) 'The ignorance of Bourbaki', *Mathematical Intelligencer*, 1992 (14): 4–13.

Mikami, Yoshio [Sanshang Yifu 三上義夫] (1933) *Zhongguo suanxue zhi tese* 中國算學之特色 *(The Character of Chinese Mathematics)*, Shanghai: Shangwu yinshuguan.

Milnor, J. (1958) 'Some consequences of a theorem of Bott', *Annals of Mathematics*, 68 (2): 444–9. Online. Available HTTP: <http://www.jstor.org/stable/1970255> (accessed 24 November 2010).

——(1963) 'Topological manifolds and smooth manifolds', *Proceedings of the International Congress of Mathematicians (Stockholm, 1962)*, Djursholm: Institut Mittag-Leffler.

——(1965) 'On the Stiefel–Whitney numbers of complex manifolds and of spin manifolds', *Topology*, 3 (3): 223–30. Online. Available HTTP: <http://www.sciencedirect.com/science/article/B6V1J-4662DC1-2/2/2d7db9cff9747a4e5cc4406e2bf70b3a> (accessed 13 December 2010).

Milnor, J. and Stasheff, J. (1974) *Characteristic Classes*, Princeton: Princeton University Press.

Mo Yueda 莫曰达 (1993) *Zhongguo tongji fazhan jian shi* 中国统计发展简史 *(A Short History of the Development of Chinese Statistics)*, Beijing: Zhongguo tongji chubanshe.

Nairn, T. (1997) *Faces of Nationalism: Janus Revisited*, London and New York: Verso.

Nasar, S. and Gruber, D. (2006) 'Manifold destiny: A legendary problem and the battle over who solved it', *The New Yorker*, August 28, 2006. Online. Available HTTP: <http://www.newyorker.com/archive/2006/08/28/060828fa_fact2#ixzz1Yskil3yp> (accessed 13 September 2011).

Nash, J. (1951) 'Non-cooperative games', *Annals of Mathematics*, 54 (2): 286–95. Online. Available HTTP: <http://www.jstor.org/stable/1969529> (accessed 15 May 2010).

Nathan, A.J. (1976) 'Policy oscillations in the People's Republic of China: A critique', *The China Quarterly*, (68): 720–33. Online. Available HTTP: <http://www.jstor.org/stable/652583> (accessed 12 February 2012).

Needham, J. (1954) *Science and Civilisation in China. Vol. I: Introductory Orientations*, Cambridge: Cambridge University Press.

Needham, J. and Wang Ling (1959) *Science and Civilisation in China. Vol. 3: Mathematics and the Sciences of the Heaven and the Earth*, Cambridge: Cambridge University Press.

Nienhauser, W.H. Jr. (ed.) (1994) *The Grand Scribe's Records, by Ssu-ma Ch'ien*. Vol. VII. *The Memoirs of Pre-Han China*. Bloomington and Indianapolis: Indiana University Press.

Oksenberg, M. (1975) 'Review: Development of Research in China', *Science*, 189 (4205): 787–8. Online. Available HTTP: <http://www.jstor.org/stable/1740203> (accessed 19 March 2009).

Ouyang Shoucheng and Peng Taoyong (2005) 'Non-modifiability of information and some problems in contemporary sciences', *Applied Geophysics*, 2 (1): 56–62. Online. Available HTTP: <http://dx.doi.org/10.1007/s11770-005-0010-z> (accessed 13 September 2011).

Parshall, K.H. (1989) 'Towards a history of nineteenth-century invariant theory', in D.E. Rowe and J. McCleary (eds) *The History of Modern Mathematics*, Vol. 1, Boston: Academic Press.

Peng Jiagui 彭家贵 and Hu Sen 湖森 (2010) 'Ti xie hou jin, wei ren shi biao – Wu Wenjun xiansheng yu Zhongguo ke da 提携后进为人师表 – 吴文俊先生与中国科大' (Carrying forward as a model teacher – Wu Wenjun and USTC), in Jiang Boju 姜伯驹, *et al.* (eds) *Wu Wenjun yu Zhongguo shuxue* 吴文俊与中国数学 *(Wu Wen-Tsun and Chinese Mathematics)*, Singapore: World Scientific.

Pepper, S. (1987) 'New directions in education', in R. MacFarquhar and J.K. Fairbank (eds) *Cambridge History of China. Vol. 14: The People's Republic, Part 1: The Emergence of Revolutionary China 1949–1965*, Cambridge: Cambridge University Press.

Pontrjagin, L.S. (1942) 'Characteristic cycles on manifolds', *C. R. Acad. Sci. URSS*, 35 (2): 34–7.

Qian Baocong 钱宝琮 (1951) 'Zhongguo gudai shuxue de weida chengjiu 中国古代数学的伟大成就' (Great achievements of ancient Chinese mathematics), *Shuxue tongbao* 数学通报, (October 1951); reprinted in *Li Yan, Qian Baocong kexue shi quanji. Vol. 9: Qian Baocong lunwen ji* 李俨、钱宝琮科学史全集。第九卷：钱宝琮论文集 *(Qian Baocong's Research Papers)*, Shenyang: Liaoning jiaoyu chubanshe.

——(1964) *Zhongguo shuxue shi* 中国数学史 *(History of Chinese Mathematics)*, Beijing: Kexue chubanshe.

Qin Jie 秦杰 and Wang Li 王黎 (2001) 'Hui dang ling jueding – shou jie guojia zui gao kexue jishu jiang chansheng shimo 会当凌绝顶 —— 首届国家最高科学技术奖产生始末' (About the first Highest National Science and Technology Award), *Xinhua News*, 19 February 2001. Online. Available HTTP: <http://news.xinhuanet.com/china/htm/20010219/371421.htm> (accessed 4 September 2011).

Qu Anjing 曲安京 (2005) 'Zhongguo shuxue shi yanjiu fanshi de zhuanhuan 中国数学史研究范式的转换' (Changing the paradigm: Research on history of mathematics in China), *Zhongguo keji shi zazhi* 中国科技史杂志 *(The Chinese Journal for the History of Science and Technology)*, (1): 50–58. Online. Available HTTP: <http://sci-cul.ihns.ac.cn/filelib//20050322001.pdf> (accessed 10 May 2008).

Qu Jingcheng 屈儆诚 and Xu Liangying 许良英 (1984) 'Guanyu woguo Wenha da geming shiqi pipan Ai'ensitan he xiangdui lun yundong de chubu kaocha 关于我国"文化大革命"时期批判爱因斯坦和相对论运动的初步考查' (An initial investigation on the critique movement of Einstein and the theory of relativity during the 'Cultural Revolution'), *Ziran bianzhengfa tongxun* 自然辩证法通讯 *(Journal of the Dialectics of Nature)*, 4 (6): 32–41.

Reid, C. (1970) *Hilbert*, Berlin: Springer.

Ren Nanheng 任南衡 and Zhang Youyu 张友余 (eds) (1995) *Zhongguo shuxue hui shiliao* 中国数学会史料 *(Historical Materials of the Chinese Mathematical Society)*. Nanjing: Jiangsu jiaoyu chubanshe.

Renmin ribao 人民日报(1957a) 'Jiangli xianjin, guwu houqi, qixiang kexue da jin jun Wo guo shou ci banfa kexue jiangjin 奖励先进，鼓舞后起，齐向科学大进军　我国首次颁发科学奖金' (Award the advanced, encourage the followers, let everyone march on science – Our country awards the first Science Prizes), *Renmin ribao* 人民日报 *(People's Daily)*, 25 January 1957: 1.

——(1957b) 'Zhongguo kexueyuan 1956 niandu kexue jiangjin (ziran kexue bufen) shenping jingguo shuoming 中国科学院1956年度科学奖金（自然科学部分）评审经过说明' (Description of the review process for the Science Prize of the Chinese Academy of Sciences (natural sciences) for the year 1956), *Renmin ribao* 人民日报 *(People's Daily)*, 25 January 1957: 7.

——(1958) 'Women de xingdong kouhao – fandui langfei, qinjian jianguo! 我们的行动口号 – 反对浪费，勤俭建国！' (Our action slogan – oppose waste, build the country efficiently!), *Renmin ribao* 人民日报 *(People's Daily)*, 2 February 1958: 1.

Richard, J.W. (2008) *Hua Loo-Keng and the Movement of Popularizing Mathematics in the People's Republic of China*, PhD Dissertation, Columbia University, New York.

Richardson, M. and Smith, P.A. (1938) 'Periodic transformations of complexes', *Annals of Mathematics*, 39 (3): 611–33. Online. Available HTTP: <http://www.jstor.org/stable/1968638> (accessed 20 April 2010).

Ritt, J.F. (1930) 'Manifolds of functions defined by systems of algebraic differential equations', *Transactions of the American Mathematical Society*, 32 (4): 569–98. Online. Available HTTP: <http://www.jstor.org/stable/1989342> (accessed 2 March 2011).

——(1932) *Differential Equations from the Algebraic Standpoint*, New York: American Mathematical Society. Online. Available HTTP: <http://www.archive.org/details/differentialequa033050mbp> (accessed 28 February 2011).

——(1950) *Differential Algebra*, American Mathematical Society.

Rosen, S. (1982) *Red Guard Factionalism and the Cultural Revolution in Guangzhou (Canton)*, Boulder: Westview.

Rowe, D.E. (2001) 'Looking back on a bestseller: Dirk Struik's *A Concise History of Mathematics*', *Notices of the AMS*, 48 (6): 590–2. Online. Available HTTP: <http://www.ams.org/notices/200106/rev-rowe.pdf> (accessed 22 September 2011).

——(2003) 'Review of Parshall, Rice: Mathematics Unbound: The Evolution of an International Mathematical Research Community', *Bulletin of the American Mathematical Society*, 40 (4): 535–42. Online. Available HTTP: <http://www.ams.org/journals/bull/2003-40-04/S0273-0979-03-00990-X/S0273-0979-03-00990-X.pdf>

Samelson, H. (1953), 'Wu, Wen-Tsun. Sur les classes caractéristiques des structures fibrées sphériques. (French)', *Mathematical Reviews Online*: MR0055691. Online. Available HTTP: <http://www.ams.org/mathscinet/> (accessed 2013-10-12).

Scheid, V. (2002) *Chinese Medicine in Contemporary China*, Durham and London: Duke University Press.

Schmalzer, S. (2006) 'Labor created humanity: Cultural Revolution science on its own terms', in J.W. Esherick, P.G. Pickowicz and A.G. Walder (eds) *The Chinese Cultural Revolution as History*, Stanford: Stanford University Press.

——(2008) *The People's Peking Man: Popular Science and Human Identity in Twentieth-Century China*, Chicago: The University of Chicago Press.

Schram, S. (1989) *The Thought of Mao Tse-Tung*, Cambridge: Cambridge University Press.

Schrijver, A. (2002) 'On the history of the transportation and maximum flow problems', *Mathematical Programming*, 91: 437–45. Online. Available HTTP: <http://homepages.cwi.nl/~lex/files/histtrpclean.pdf>.

Scott, J.F. (1958) *A History of Mathematics*, London: Taylor and Francis.

Segal, S.L. (2002) 'War, refugees, and the creation of an international mathematical community', in K.H. Parshall and A.C. Rice (eds) *Mathematics Unbound: The Evolution of an International Mathematical Research Community, 1800–1945*, Providence: American Mathematical Society.

Seneta, E. (2004) 'Mathematics, religion, and Marxism in the Soviet Union in the 1930s', *Historia Mathematica*, 31: 337–67.

Shapiro, A. (1957) 'Obstructions to the imbedding of a complex in a Euclidean space. I. The first obstruction', *Annals of Mathematics*, 66 (2): 256–69. Online. Available HTTP: <http://www.jstor.org/stable/1969998> (accessed 20 April 2010).

Shen Kangshen 沈康身, Crossley, J.N. and Lun, A.W.-C. (1999) *The Nine Chapters on the Mathematical Art. Companion and Commentary*, Oxford and Beijing: Oxford University Press-Science Press.

Shen Shihao 沈世豪 (1997) *Chen Jingrun zhuan* 陈景润传 (Biography of Chen Jingrun), Xiamen: Xiamen daxue chubanshe. Online. Available HTTP: <http://www.chinaunix.net/old_jh/31/414198.html> (accessed 30 December 2013).

Shi Jihuai 史济怀 (ed.) (2008) *Zhongguo kexue jishu daxue shuxue wu shi nian* 中国科学技术大学数学五十年 *(Fifty Years of Mathematics at the University of Science and Technology of China)*. Hefei: Zhongguo kexue jishu daxue chubanshe.

Short, P. (2004) *Mao: A Life*, London: J. Murray.

Shu Jin 舒进 (1974a) 'Xuexi Zhongguo shuxue shi ziliao zhaji – *Da Ming li* chansheng de douzheng yu Zu Chongzhi de shuxue chengjiu 学习中国数学史资料札记 – 《大明历》产生时的斗争与祖冲之的数学成就' (Notes from studying materials for the history of Chinese mathematics – The struggle during the formation of the *Calendar of Great Clarity* and the mathematical achievements of Zu Chongzhi), *Shuxue xuebao* 数学学报 *(Acta Mathematica Sinica)*, 17 (3): 153–5.

——(1974b) 'Rujia de fandong sixiang dui wo guo gudai shuxue de fazhan qi zhe zu'ai he pohuai de zuoyong 儒家的反动思想对我国古代数学的发展起着阻碍和破坏的作用' (Confucian reactionary thought obstructed and damaged the development of our ancient mathematics), *Shuxue xuebao* 数学学报 *(Acta Mathematica Sinica)*, 17 (4): 227–33.

Shu Qun 舒群 (1974) 'Weida de jinbu kexuejia Zu Chongzhi', *Shuxue de shijian yu renshi* 数学的实践与认识 *(Mathematics in Practice and Theory)*, 3 (4): 1–6.

Shuxue jinzhan 数学进展 (1955) 'Shuxuejie dongtai 数学界动态'(News from mathematical circles), *Shuxue jinzhan* 数学进展 *(Advances in Mathematics)*, 1 (3): 404.

——数学进展 (1956) 'Xiaoxi shu ze 消息数则' (A few notices), *Shuxue jinzhan* 数学进展 *(Advances in Mathematics)*, 2 (2): 308–9.

——数学进展 (1957) 'Beijing shuxuejie xueshu huodong dongtai 北京数学界学术活动动态' (Academic activities in Beijing mathematical circles), *Shuxue jinzhan* 数学进展 *(Advances in Mathematics)*, 3 (3): 492–3.

Skinner, G.W. and Winckler, A.E. (1969) 'Compliance succession in rural Communist China: A cyclical theory', in A. Etzioni (ed.) *Complex Organizations: A Sociological Reader*, 2nd edn, New York: Holt, Rinehart, and Winston.

Smith, A.D. (1981) *The Ethnic Revival*, Cambridge: Cambridge University Press.

Smith, P.A. (1941) 'Periodic and nearly periodic transformations', in R.L. Wilder and W.L. Ayres (eds) *Lectures in Topology*, Ann Arbor: University of Michigan Press.

——(1956) *J.F. Ritt: A Biographical Memoir*, National Academy of Sciences.

Stasheff, J. (1982), 'Wu, Wen Tsün [Wu, Wen Jun] deRham–Sullivan measure of spaces and its calculability', *Mathematical Reviews Online*: MR609563. Online. Available HTTP: <http://www.ams.org/mathscinet/> (accessed 23 November 2013).

Stavis, B. (1975) 'Review: Research and Revolution. Science Policy and Societal Change in China', *The Journal of Asian Studies*, 35 (1): 136–8. Online. Available HTTP: <http://www.jstor.org/stable/2054054> (accessed 19 March 2009).

Steenrod, N.E. (1947) 'Products of cocycles and extensions of mappings', *Annals of Mathematics*, 48 (2): 290–320. Online. Available HTTP: <http://www.jstor.org/stable/1969172> (accessed 5 March 2010).

——[1951] (1974) *The Topology of Fibre Bundles*, 2nd edn, Princeton: Princeton University Press.

Stiefel, E. (1935) 'Richtungsfelder und Fernparallelismus in n-dimensionalen Mannigfaltigkeiten', *Commentarii Mathematici Helvetici*, 22: 305–53.

Stone, M.H. (1961) 'Mathematics, 1949–1960', in S.H. Gould (ed.) *Sciences in Communist China*, Washington, DC: American Association for the Advancement of Science.

Struik, D.J. (1948) *A Concise History of Mathematics*, New York: Dover.

Su Yang (2006) 'Mass killings in the Cultural Revolution: A study of three provinces', in J.W. Esherick, P.G. Pickowicz and A.G. Walder (eds) *The Chinese Cultural Revolution as History*, Stanford: Stanford University Press.

Sullivan, D. (1973) 'Differential forms and the topology of manifolds', presented at *Proceedings of the International Conference on Manifolds*, Tokyo.

Sun Wenhua 孙文晔 (2008) 'Gedebahe "Caixiang" bao chun 哥德巴赫"猜想"报春' (Goldbach 'Conjecture' announced spring), *Beijing ribao* 北京日报 *(Beijing Daily)*, 25 December 2008. Online. Available HTTP: <http://www2.cas.cn/html/Dir/2008/12/23/16/34/33.htm> (accessed 22 September 2011)

Sun Yifeng 孙以丰 (2010) 'Huiyi yi ge tuopu xiaozu de er san shi 回忆一个拓扑小组的二三事' (Remembering a few stories from a topology group), in Jiang Boju 姜伯驹, *et al.* (eds) *Wu Wenjun yu Zhongguo shuxue* 吴文俊与中国数学 *(Wu Wen-Tsun and Chinese Mathematics)*, Singapore: Global Publishing.

Sun Yongsheng 孙永生 (1972) 'Guanyu gongli fangfa 关于公理方法' (About the Axiomatic Method), *Shuxue de shijian yu renshi* 数学的实践与认识 *(Mathematics in Practice and Theory)*, 2 (4): 51–8.

Suttmeier, R.P. (1970) 'Party views of science: The record from the first decade', *The China Quarterly*, 44: 146–68. Online. Available HTTP: <http://www.jstor.org/stable/651958> (accessed 12 May 2009).

——(1974) *Research and Revolution: Science Policy and Societal Change in China*, Lexington, MA: Lexington Books.

——(1980) *Science, Technology and China's Drive for Modernization*, Stanford: Hoover Institution Press.

Tarski, A. (1957) *A Decision Method for Elementary Algebra and Geometry*, 2nd edn, Santa Monica: RAND Corporation. Online. Available HTTP: <http://www.rand.org/pubs/reports/2008/R109.pdf> (accessed 12 March 2009).

——(1959) 'What is elementary geometry?', in L. Henkin, P. Suppes and A. Tarski (eds) *The Axiomatic Method*, Amsterdam: North-Holland Publishing Company.

Taylor, K. (2005) *Chinese Medicine in Early Communist China, 1945–63*, London: Routledge.

Teiwes, F.C. (1987). 'Establishment and consolidation of the new regime', in R. MacFarquhar and J.K. Fairbank (eds) *Cambridge History of China. Vol. 14: The People's Republic, Part 1: The Emergence of Revolutionary China 1949–1965*, Cambridge: Cambridge University Press.

Teiwes, F.C. and Sun, W. (1996) *The Tragedy of Lin Biao: Riding the Tiger during the Cultural Revolution*, Honolulu: University of Hawaii Press.

——(2007) *The End of the Maoist Era: Chinese Politics During the Twilight of the Cultural Revolution, 1972–1976*, Armonk and London: M. E. Sharpe.

The Abel Committee (2004), 'The Abel Prize Laureates 2004: Citation and biography'. Online. Available HTTP: <http://www.abelprize.no/c53865/binfil/download.php?tid=53806> (accessed 13 October 2013).

The Shaw Prize Secretariat (2006), 'The Shaw Laureates in Mathematical Sciences 2006'. Online. Available HTTP: <http://www.shawprize.org/en/shaw.php?tmp=3&t woid=51&threeid=63&fourid=103> (accessed 12 January 2013).

Thiele, R. (2005) 'Hilbert and his twenty-four problems', in M. Kinyon and G. van Brummelen (eds) *Mathematics and the Historian's Craft: The Kenneth O. May Lectures*, Springer. Online. Available HTTP: <http://www.springerlink.com/content/ w8kg0666281823nw/> (accessed 22 September 2011)

Thom, R. (1950) 'Classes caractéristiques et i-carrés', *C. R. Acad. Sci. Paris*, 230: 427–9.

——(1952a) 'Espaces fibrés en sphères et carrés de Steenrod', *Annales scientifiques de l' Ecole normale supérieure. (3)*, 69: 109–82.

——(1952b) 'Une théorie intrinsèque des puissances de Steenrod', *Colloque de Topologie* Vol. 1951, Strasbourg: Bibliothèque nationale et universitaire de Strasbourg.

——(1990) 'Problèmes rencontrés dans mon parcours mathématique: un bilan', *Publications mathématiques de l'I.H.É.S*, 70: 199–214. Online. Available HTTP: <http://archive.numdam.org/article/PMIHES_1989__70__199_0.pdf> (accessed 13 September 2011).

Tian Miao 田淼 (2000) 'Chen Xingshen caifang lu 陈省身采访录' (Interview with S.S. Chern), *Zhongguo keji shiliao* 中国科技史料 *(Historical Materials of Chinese Science and Technology)*, 21 (2): 117–27.

Toepell, M.-M. (1986) 'On the origins of David Hilbert's *Grundlagen der Geometrie*', *Archive for History of Exact Sciences*, 35 (4): 329–44.

Tonelli, L. (1929) 'Report on the 1928 International Congress of Mathematicians', *Bulletin of the American Mathematical Society*, 35 (2): 201–4. Online. Available HTTP: <http://www.ams.org/journals/bull/1929-35-02/S0002-9904-1929-04700-1/ S0002-9904-1929-04700-1.pdf> (accessed 22 September 2011).

U Ven'-tszyun [吴文俊 Wu Wen-Tsun] (1963) 'O beskoalitsionnych igrakh s ogranichenniyami na oblasti izmeneniya strategiy' (On non-cooperative games with restricted domains of activities), in N.N. Vorob'ev (ed.) *Beskonechnye antagonisticheskie igry (Infinite Antagonistic Games)*, Moscow: Fizmatgiz.

van der Waerden, B.L. (1939) *Einführung in die algebraische Geometrie*, Berlin: Springer.

van Kampen, E. (1932) 'Komplexe in euklidischen Räumen', *Abhandlungen aus dem Mathematischen Seminar der Universität Hamburg*, 9: 72–8.

Vetter, Q. (1929) 'Mezinárodní kongres matematiků v Bologni' (International Congress of Mathematicians in Bologna), *Časopis pro pěstování matematiky a fyziky*, 58 (1–2): 184–5. Online. Available HTTP: <http://dml.cz/bitstream/handle/10338.dmlcz/108930/ CasPestMatFys_058-1929-1_32.pdf> (accessed 22 September 2011).

von Neumann, J. (1928) 'Zur Theorie der Gesellschaftsspiele', *Mathematische Annalen*, 100 (1): 295–320.

Vorob'ev, N.N. (1970) 'The present state of the theory of games', *Russian Mathematical Surveys*, 25 (2): 77. Online. Available HTTP: <http://iopscience.iop.org/0036-0279/25/2/R04> (accessed 24 May 2011).

Wagner, D.B. (1978) 'Liu Hui and Zu Gengzhi on the volume of a sphere', *Chinese Science*, 3: 59–79. Online. Available HTTP: <http://www.staff.hum.ku.dk/dbwagner/ SPHERE/SPHERE.html> (accessed 12 June 2009).

——(1979) 'An ancient Chinese derivation of the volume of a pyramid: Liu Hui, third century A.D.', *Historia Mathematica*, 6: 164–88. Online. Available HTTP: <http:// www.staff.hum.ku.dk/dbwagner/Pyramid/Pyramid.html> (accessed 12 June 2009).

Walder, A.G. (1991) 'Cultural Revolution radicalism: Variations on a Stalinist theme', in W.A. Joseph, C.P.W. Wong and D. Zweig (eds) *New Perspectives on the Cultural Revolution*, Cambridge, MA: The Council on East Asian Studies.

——(2002) 'Beijing Red Guard factionalism: Social interpretations reconsidered', *Journal of Asian Studies*, 61 (2): 437–71. Online. Available HTTP: <http://www.jstor.org/stable/2700297>

Wan Zhexian 万哲先 (1955) 'Henglixi Geleier jiaoshou fanghua jianbao 亨利希格雷尔教授放花简报' (A brief report about the visit of Professor Heinrich Grell to China), *Shuxue jinzhan* 数学进展 *(Advances in Mathematics)*, 1 (3): 403.

——(1997) *Wan Zhexian shuxue kepu wenxuan* 万哲先数学科普文选 *(Selection of Mathematics Popularization Articles by Wan Zhexian)*, Shijiazhuang: Hebei kexue jishu chubanshe.

Wang Dongming 王东明, *et al.* (2010) 'Shuxue jixiehua fazhan huigu 数学机械化发展回顾' (Review of the development of mechanization of mathematics), in Jiang Boju 姜伯驹, *et al.* (eds) *Wu Wenjun yu Zhongguo shuxue* 吴文俊与中国数学 *(Wu Wen-Tsun and Chinese Mathematics)*, Singapore: Global Publishing.

Wang Hao (1960) 'Toward mechanical mathematics', *IBM Journal*, 1960 (1): 2–22.

Wang Lili 王丽丽 and Li Xiaoning 李小凝 (1998) *Chen Jingrun zhuan* 陈景润传 *(Chen Jingrun's Biography)*, Beijing: Xinhua chubanshe.

Wang Qian Guozhong 王钱国忠 (2000) 'Li Yuese yanjiu de huigu yu zhanwang 李约瑟研究的回顾与瞻望' (Review and new perspectives on Needham studies), in 李约瑟文献中心 Li Yuese wenxian zhongxin (ed.) *Li Yuese yanjiu* 李约瑟研究就, Vol. 1, Shanghai: Shanghai kexue puji chubanshe.

Wang Yangzong 王扬宗 and Cao Xiaoye 曹效业 (eds) (2010). *Zhongguo Kexueyuan yuanshu danwei jian shi* 中国科学院院属单位简史 *(A Short History of the Units under Chinese Academy of Science)*. Vol. 1. Beijing: Kexue chubanshe.

Wang Yao 王尧 (1975) 'Tan tan Zhongguo shuxue shi 谈谈中国数学史' (About the history of Chinese mathematics), *Shuxue xuebao* 数学学报 *(Acta Mathematica Sinica)*, 18 (3): 157–61.

Wang Yuan (1999) *Hua Loo-Keng*, Singapore: Springer.

Wei Chunjuan, N. and Brock, D.E. (2013) *Mr. Science and Chairman Mao's Cultural Revolution: Science and Technology in Modern China*, Plymouth: Lexington.

Weigelin-Schwiedrzik, S. (1996) 'On Shi and Lun: Toward a typology of historiography in the PRC', *History and Theory*, 35 (4): 74–95. Online. Available HTTP: <http://www.jstor.org/stable/2505445> (accessed 20 March 2011).

Weil, A. (1992) *The Apprenticeship of a Mathematician*, Boston: Birkhäuser.

Whitney, H. (1935) 'Sphere spaces', *Proceedings of the National Academy of Sciences*, 21: 464–8.

——(1941) 'On the topology of differentiable manifolds', in R.L. Wilder and W.L. Ayres (eds) *Lectures in Topology*, Ann Arbor: University of Michigan Press.

——(1944a) 'The self-intersections of a smooth n-manifold in 2n-space', *Annals of Mathematics*, 45: 220–46.

——(1944b) 'The singularities of a smooth n-manifold in (2n−1)-space', *Annals of Mathematics*, 45: 247–93.

——(1949), 'Wu, Wen-tsun: On the product of sphere bundles and the duality theorem modulo two', *Mathematical Reviews Online*: MR0026794. Online. Available HTTP: <http://www.ams.org/mathscinet/> (accessed 2013-10-12).

Wilder, R.L. and Ayres, W.L. (eds) (1941). *Lectures in Topology*. Ann Arbor: University of Michigan Press.

Wilson, W.S., *et al.* (1996) 'Wei-Liang Chow', *Notices of the AMS*, 43 (10): 1117–24. Online. Available HTTP: <http://www.ams.org/notices/199610/chow.pdf> (accessed 29 October 2011).

Wu Maogui 吴茂贵 and Xu Kang 许康 (1987) 'Qin Jiushao "san xie qiu ji" gongshi gu zheng zai tan 秦九韶"三邪求积"公式古证再探' (Further examination of ancient proofs of Qin Jiushao's formula for "finding the area of three skewed"), presented at *International Symposium on the 740th anniversary of Qin Jiushao's 'Mathematical Treatise in Nine Sections'*, Beijing. (unpublished). Quoted from NRI Offprint Collection, Mathematics XIV.

Wu Wen-Tsun 吴文俊 (1947) 'Note sur les produits essentiels symétriques des espaces topologiques' (A note on essential symmetric products of topological spaces), *C. R. Acad. Sci. Paris*, 224: 1139–41. Online. Available HTTP: <http://gallica.bnf.fr/ark:/12148/bpt6k3176m/f1143.image.r=comptes%20rendus%20hebdomadaires.langEN> (accessed 13 March 2010).

——(1948a) 'On the product of sphere bundles and the duality theorem modulo two', *Annals of Mathematics*, 49 (3): 641–53. Online. Available HTTP: <http://www.jstor.org/stable/1969049> (accessed 12 March 2010).

——(1948b) 'Sur l'existence d'un champ d'éléments de contact ou d'une structure complexe sur une sphère' (On the existence of a field of elements of contact or of a complex structure on a sphere), *C. R. Acad. Sci. Paris*, 226: 2117–9. Online. Available HTTP: <http://gallica.bnf.fr/ark:/12148/bpt6k31787/f2117.image.r=comptes%20rendus%20hebdomadaires.langEN> (accessed 13 March 2010).

——(1950a) 'Classes caractéristiques et i-carrés d'une variété' (Characteristic classes and i-squares of a manifold), *C. R. Acad. Sci. Paris*, 230: 508–11. Online. Available HTTP: <http://gallica.bnf.fr/ark:/12148/bpt6k3182n/f508.image.r=comptes%20rendus%20hebdomadaires.langEN> (accessed 13 March 2010).

——(1950b) 'Les i-carrés dans une variété grassmannienne' (The i-squares of a Grassmanian manifold), *C. R. Acad. Sci. Paris*, 230: 918–20. Online. Available HTTP: <http://gallica.bnf.fr/ark:/12148/bpt6k3182n/f918.image.r=comptes%20rendus%20hebdomadaires.langEN> (accessed 13 March 2010).

——(1951) 'Faguo shuxue xin pai – Buerbaji pai 法国数学新派 —— 布尔巴基派' (A new school in French mathematics – the Bourbaki school), *Kexue tongbao* 科学通报, 2 (4): 415.

——(1952) 'Sur les puissances de Steenrod' (On Steenrod powers), presented at *Colloque de Topologie* Strasbourg. Bibliothèque Nationale et Universitaire de Strasbourg.

——(1953a) 'On squares in Grassmannian manifolds', *Scientia Sinica*, 2: 91–115. Online. Available HTTP: <http://www.scichina.com:8081/sciAe/fileup/PDF/53ya0091.pdf> (accessed 19 February 2011).

——(1953b) 'Youxian ke peifen kongjian de xin tuopu bubianliang 有限可剖分空间的新拓扑不变量' (Topological invariants of new type of finite polyhedrons), *Shuxue xuebao* 数学学报 *(Acta Mathematica Sinica)*, 3 (4): 261–90.

——(1953c) 'Lun Pontryagin shixinglei 论понтрягин示性类 I' (On Pontryagin classes I), *Shuxue xuebao* 数学学报 *(Acta Mathematica Sinica)*, 3 (4): 291–315.

——(1954a) 'On Pontrjagin classes. I', *Scientia Sinica*, 3: 353–67. Online. Available HTTP: <http://www.scichina.com:8081/sciAe/fileup/PDF/54ya0353.pdf> (accessed 19 February 2011).

——(1954b) 'Lun Pontryagin shixinglei 论понтрягин示性类 II' (On Pontryagin classes II), *Shuxue xuebao* 数学学报 *(Acta Mathematica Sinica)*, 4: 171–99.

——(1954c) 'Lun Pontryagin shixinglei 论понтрягин示性类 III' (On Pontryagin classes III), *Shuxue xuebao* 数学学报 *(Acta Mathematica Sinica)*, 4: 323–46.

——(1955a) 'Bu neng wangji de yi jian shi 不能忘记的一件事' (An event I cannot forget), *Shuxue tongbao* 数学通报, 1955 (7): 4.

——(1955b) 'On Pontrjagin classes II', *Scientia Sinica*, 4: 455–90.

——(1955c) 'Lun Pontryagin shixinglei 论понтрягин示性类 IV' (On Pontryagin classes IV), *Shuxue xuebao* 数学学报 *(Acta Mathematica Sinica)*, 5 (1): 37–63.

——(1955d) 'Fuhexing zai Ou shi kongjian zhong de shixian wenti 复合形在欧氏空间中的实现问题 I' (On the realization of complexes in Euclidean spaces I), *Shuxue xuebao* 数学学报 *(Acta Mathematica Sinica)*, 5 (4): 505–52.

——(1956) 'Shuxue de fangfa lun 数学的方法论' (Methodology of mathematics), *Ziran bianzhengfa yanjiu tongxun* 自然辩证法研究通讯 *(Research Notices about the Dialectics of Nature)*, 0 (0): 118, 91.

——(1957a) 'On the F(p) classes of a topological space', *Science Record*, 1: 377–80.

——(1957b) 'Fuhexing zai Ou shi kongjian zhong de shixian wenti 复合形在欧氏空间中的实现问题 II' (On the realization of complexes in Euclidean spaces II), *Shuxue xuebao* 数学学报 *(Acta Mathematica Sinica)*, 7 (1): 79–101.

——(1957c) 'On the relations between Smith operations and Steenrod powers', *Fundamenta Mathematicae*, 44: 262–9.

——(1958a) 'On the isotopy of Cr-manifolds of dimension (n) in Euclidean (2n+1)-space', *Science Record*, 2: 271–5.

——(1958b) 'Fuhexing zai Ou shi kongjian zhong de shixian wenti 复合形在欧氏空间中的实现问题 III' (On the realization of complexes in Euclidean spaces III), *Shuxue xuebao* 数学学报 *(Acta Mathematica Sinica)*, 8 (1): 79–94.

——(1959a) 'Boyi lun zatan: (yi) er ren boyi 博弈论杂谈:(一)二人博弈' (On game theory: (1) Two-person games), *Shuxue tongbao* 数学通报, 1959 (10): 339–44.

——(1959b) 'On the isotopy of a finite complex in a Euclidean space I, II', *Science Record*, 3: 342–51.

——(1959c) 'A remark on the fundamental theorem in the theory of games', *Science Record*, 3: 229–33.

——(1960a) 'On the isotopy of a complex in a Euclidean space I', *Scientia Sinica*, 9 (1): 21–46.

——(1960b) 'Shuxue zai guomin jingji de yingyong 数学在国民经济的应用' (Applications of mathematics in the national economy), *Shuxue tongbao* 数学通报, 1960 (4): 25–7.

——(1961a) 'Huodong shou xianzhi xia de fei xiezuo duice 活动受限制下的非协作对策' (On non-cooperative games with restricted domains of activities), *Shuxue xuebao* 数学学报 *(Acta Mathematica Sinica)*, 11 (1): 47–62.

——(1961b) 'Guanyu Leray de yi ge dingli 关于Leray的一个定理' (On a theorem of Leray), *Shuxue xuebao* 数学学报 *(Acta Mathematica Sinica)*, 11 (4): 348–56.

——(1962) *Lixue zai jihe zhong de yixie yingyong* 力学在几何中的一些应用 *(Some Applications of Mechanics in Geometry)*, Beijing: Renmin jiaoyu chubanshe.

——(1964a) 'A theorem on immersion', *Scientia Sinica*, 13 (1): 160.

——(1964b) 'On the immersion of C∞-3 manifolds in a Euclidean space', *Scientia Sinica*, 13 (2): 335–6.

——(1965a) *A Theory of Imbedding, Immersion, and Isotopy of Polytopes in a Euclidean Space*, Peking: Science Press.

——(1965b) 'Daishu cu shang de Chen Xingshen shixing lei 代数簇上的陈省身示性类' (The Chern characteristic classes on an algebraic variety), *Shuxue jinzhan* 数学进展 *(Advances in Mathematics)*, 8 (4): 395–401.

——(1965c) 'Juyou dui'ou youli fen'ge de daishu cu 具有对偶有理分割的代数簇' (Algebraic varieties with dual rational dissections), *Shuxue jinzhan* 数学进展 *(Advances in Mathematics)*, 8 (4): 402–9.

——(1973) 'Jicheng dianlu sheji zhong de yi ge shuxue wenti 集成电路设计中的一个数学问题' (A mathematical problem in the design of integrated circuits), *Shuxue de shijian yu renshi* 数学的实践与认识 *(Mathematics in Practice and Theory)*, 2 (1): 20–40.

——(1974) 'Sk xing qidian suoshu de tongtiao lei Sk型奇点所属的同调类' (Homology classes that contain singularities of type Sk), *Shuxue xuebao* 数学学报 *(Acta Mathematica Sinica)*, 17 (1): 28–37.

——(1975a) 'Daishu tuopu I^* hanzi lun – qixing kongjian de shi tuopu 代数拓扑 I *函子论 —— 齐性空间的实拓扑' (Theory of I^*-functor in algebraic topology – Real topology of homogeneous spaces), *Shuxue xuebao* 数学学报 *(Acta Mathematica Sinica)*, 18 (3): 162–72.

——(1975b) 'Daishu tuopu de yi ge xin hanzi 代数拓扑的一个新函子' (A new functor in algebraic topology), *Kexue tongbao* 科学通报, 20 (10): 311–12.

——(1975c) 'Theory of I^*-functor in algebraic topology – real topology of fibre squares', *Scientia Sinica*, 18 (4): 464–82.

——(1976) 'Wenhua da geming wei shuxue yanjiu kaipi le guangkuo de qiantu 文化大革命为数学研究开辟了广阔的前途' (The Cultural Revolution opened up a broad future for mathematical research), *Yingyong shuxue xuebao* 应用数学学报 *(Acta Mathematica Applicata)*, 1 (2): 13–16.

——(1977a) 'Daishu tuopu I^* hanzi lun – fuxing shang I^* hanzi de juti jisuan yu gongli xitong 代数拓扑 I^* 函子论 —— 复形上 $I{\sim}^*$ 函子的具体计算与公理系统' (Theory of I^*-functor in algebraic topology – Effective calculation and axiomatization of I^*-functor on complexes), *Shuxue xuebao* 数学学报 *(Acta Mathematica Sinica)*, 19 (3): 195–209.

——(1977b) 'Chudeng jihe panding wenti yu jixiehua zhengming 初等几何判定问题与机械化证明' (On the decision problem and the mechanization of theorem-proving in elementary geometry), *Zhongguo kexue* 中国科学 *(Scientia Sinica)*, 1977 (6): 507–16.

——(1978a) 'Chu ru xiang bu yuanli 出入相补原理' (Out–In complementary principle), in Ziran kexue shi yanjiusuo 自然科学史研究所 (ed.) *Zhongguo gudai keji chengjiu* 中国古代科技成就 *(Achievements of Ancient Chinese Science and Technology)*, Beijing: Zhongguo qingnian chubanshe; reprinted in Wu Wen-Tsun (1996) *Wu Wenjun lun shuxue jixiehua* 吴文俊论数学机械化, Jinan: Shandong jiaoyu chubanshe, 170–88.

——(1978b) 'Chudeng weifen jihe de jixiehua zhengming 初等微分几何的机械化证明' (On the mechanization of theorem-proving in elementary differential geometry), *Kexue tongbao* 科学通报, 23 (9):523–534.

——(1978c) *Kepouxing zai Ou shi kongjian zhong de shixian wenti,* 可剖形在欧氏空间中的实现问题 *(Problems of Realization of Polytopes in Euclidean Space)*, Beijing: Kexue chubanshe.

——(1978d) 'On the decision problem and mechanization of theorem-proving in elementary geometry', *Scientia Sinica*, 21 (2): 159–72.

——(1978e) 'Shuxue de jixiehua wenti 数学的机械化问题' (The question of mechanization of mathematics), *Ziran bianzhengfa tongxun* 自然辩证法通讯 *(Journal of the Dialectics of Nature)*; reprinted in Wu Wen-Tsun (1996) *Wu Wenjun lun shuxue jixiehua* 吴文俊论数学机械化, Jinan: Shandong jiaoyu chubanshe, 374.

——(1978f) 'Shuxue yu si ge xiandai hua 数学与四个现代化' (Mathematics and the Four Modernizations), *Kexuejia tan shu li hua (Scientists Talk about Mathematics, Physics and Chemistry)* 科学家啁数理化, Beijing: Zhongguo shaonian ertong chubanshe; reprinted in Wu Wen-Tsun (1996) *Wu Wenjun lun shuxue jixiehua* 吴文俊论数学机械化, Jinan: Shandong jiaoyu chubanshe, 24–7.

——(1979a) 'Chudeng weifen jihe de jixiehua zhengming 初等微分几何的机械化证明' (On the mechanization of theorem-proving in elementary differential geometry), *Zhongguo kexue* 中国科学 *(Scientia Sinica)*, 1979 (S1): 94–102.

——(1979b) 'Tuopu zhong de liangdu yu neng jisuan xing 拓扑中的量度与能计算性' (Measure and calculability in topology), *Beijing shuxue hui 1979 nian nianhui lunwen zhaiyao huibian* 北京数学会1979年年会论文摘要汇编, Beijing: Beijing shuxue hui.

——(1980) 'Shuxue de jixiehua 数学的机械化' (Mechanization of mathematics), *Baike zhishi* 百科知识 *(Encyclopaedic Knowledge)*, 1980 (3): 41–4; reprinted in Wu Wen-Tsun (1996) *Wu Wenjun lun shuxue jixiehua* 吴文俊论数学机械化, Jinan: Shandong jiaoyu chubanshe, 358–65.

——(1981a) 'Gu zheng fuyuan de yuanze 古证复原的原则' (Principles of reconstruction of proofs), *Shijie kexue* 世界科学 *(World Science)*, 1981 (10): 56.

——(1981b) 'Mechanical theorem proving in elementary geometry and differential geometry', in *Proceedings of the 1980 Beijing Symposium on Differential Geometry and Differential Equations*, Beijing. Science Press.

——(1981c) 'Shuxue zhong de gonglihua yu jixiehua sixiang 数学中的公理化与机械化思想' (Axiomatisation and mechanization thought in mathematics), *Kecheng, jiaocai, jiaofa* 课程、教材、教法 *(Courses, Teaching materials, Didactics)*, 1 (1): 28–9; reprinted in Wu Wen-Tsun (1996) *Wu Wenjun lun shuxue jixiehua* 文俊论数学机械化, Jinan: Shandong jiaoyu chubanshe, 375–7.

——(1982a) 'Hai dao suan jing gu zheng tan yuan 《海岛算经》古证探源' (Reconstructions of ancient proofs for the *Sea Island Mathematical Classic*), in 吴文俊 Wu Wen-Tsun (ed.) *Jiu zhang suan shu yu Liu Hui* 《九章算术》与刘徽 *(Nine Chapters and Liu Hui)*, Beijing: Beijing shifan daxue chubanshe.

——(1982b) 'Toward mechanization of geometry – Some comments on Hilbert's *Grundlagen der Geometrie*', *Acta Mathematica Scientia*, 2: 125–38.

——(1982c) 'Wo guo gudai cewang zhi xue chongcha lilun pingjie jian ping shuxe shi yanjiu zhong mouxie fangfa wenti 我国古代测望之学重差理论评介兼评数学史研究中某些方法问题' (On the Double-Difference Theory in ancient Chinese surveying, with comments on certain methodological questions in the history of mathematics), *Keji shi wen ji* 科技史文集, 8: 10–31.

——(1984a) 'Basic principles of mechanical theorem proving in elementary geometries', *Journal of Systems Science and Mathematical Sciences*, 3 (3): 207–35.

——(1984b) *Jihe dingli jiqi zhengming de jiben yuanli (chudeng jihe bufen)*, 几何定理机器证明的基本原理-(初等几何部分) *(Basic Principles of Mechanical Theorem-Proving in Geometries (Elementary Geometry))*, Beijing: Kexue chubanshe.

——(1984c) 'Zai Jiaoyu bu zhuban de quanguo gaoxiao zhong wai shuxue shi jiangxi ban kaixue dianli shang de jianghua 在教育部主办的全国高校中外数学史啁习班开学典礼上的啁话' (Talks at the opening ceremony of the National Higher Education Seminar on the history of Chinese and foreign mathematics, organized by the Ministry of Education), *Zhongguo shuxue shi lunwen ji* 中国数学史论文集, 2 (1986): 1–8.

——(1985a) 'Fuxing gouzaoxing de shuxue 复兴构造性的数学' (Let us revive constructive mathematics), *Shuxue jinzhan* 数学进展 *(Advances in Mathematics)*, 14 (4): 334–9.

——(1985b) 'Guanyu daishu fangcheng zu de lingdian – Ritt yuanli de yi ge yingyong 关于代数方程组的零点 —— Ritt原理的一个应用' (On zeros of algebraic equations – an application of the Ritt Principle), *Kexue tongbao* 科学通报, 30 (12).

——(1986a) 'On zeros of algebraic equations – an application of the Ritt Principle', *Chinese Science Bulletin*, 31 (1): 1–5.

——(1986b) 'Recent studies of the history of Chinese mathematics', in A.M. Gleason (ed.) *International Congress of Mathematicians 1986*, Vol. 2, Berkeley: University of California. Online. Available HTTP: <http://www.mathunion.org/ICM/ICM1986.2/ Main/icm1986.2.1657.1667.ocr.pdf> (accessed 24 July 2011).

——(1986c) *Wu Wenjun wen ji*, 吴文俊文集 *(Collection of Wu Wen-Tsun's Papers)*, Jinan: Shandong jiaoyu chubanshe.

——(1986d) 'Jiefangchengqi huo Solver ruanjian xitong gaishu 〈解方程器〉或 〈Solver〉软件系统概述' (A general description of the SOLVER software system), *Shuxue de shijian yu renshi* 数学的实践与认识 *(Mathematics in Practice and Theory)*, 15 (2): 32–9.

——(1987a) 'Cong *Shu shu jiu zhang* kan Zhongguo chuantong shuxue gouzaoxing yu jixiehua de tese 从《数书九章》看中国传统数学构造性与机械化的特色' (On the constructiveness and mechanisation of Chinese traditional mathematics seen in the *Mathematical Treatise in Nine Sections*), in 吴文俊 Wu Wen-Tsun (ed.) *Qin Jiushao yu Shu shu jiu zhang* 秦九韶与《述书九章》, Beijing: Beijing shifan daxue chubanshe.

——(1987b) 'Dui Zhongguo chuantong shuxue de zai renshi 对中国传统数学的再认识' (A new understanding of Chinese traditional mathematics), *Baike zhishi* 百科知识 *(Encyclopaedic Knowledge)*, 1987 (7–8): 48–51, 43–6; reprinted in Wu Wen-Tsun (1996) *Wu Wenjun lun shuxue jixiehua* 文俊论数学机械化, Jinan: Shandong jiaoyu chubanshe, 30–44.

——(1987c) *Rational Homotopy Type: A Constructive Study via the Theory of the I*-measure*, New York: Springer.

——(1987d) 'A mechanization method of geometry and its applications – I. Distances, areas, and volumes in Euclidean and Non-Euclidean geometries', *Chinese Science Bulletin*, 32 (7): 436–40.

——(1987e) 'A mechanization method of geometry and its applications – II. Curve pairs of Bertrand type', *Chinese Science Bulletin*, 32 (9): 585–8.

——(1987f) 'Mechanical derivation of Newton's gravitational laws from Kepler's laws', *MM Research Preprints*, 1: 53–61.

——(1987g) 'A constructive theory of differential algebraic geometry based on works of J. F. Ritt with particular applications to mechanical theorem-proving of differential geometries', *Differential Geometry and Differential Equations (Shanghai, 1985)*, Berlin: Springer. Online. Available HTTP: <http://dx.doi.org/10.1007/ BFb0077689> (accessed 22 September 2011)

——(1988a) 'Shuxue 数学' (Mathematics), *Zhongguo da baike quanshu* 中国大百科全书 *(Great Chinese Encyclopedia)*, Beijing: Zhongguo da baike quanshu chubanshe.

——(1988b) 'A mechanization method of geometry and its applications III. Mechanical proving of polynomial inequalities and equation-solving', *Systems Science and Mathematical Sciences*, 1 (1): 1–17. Online. Available HTTP: <http://www.sysmath. com/xtkxysx_en/qikan/manage/wenzhang/jssc-88-1(1)-001.pdf> (accessed 16 June 2011).

——(1989a) 'Zhongyang yanjiuyuan shuxue yanjiusuo yi nian de huiyi 中央研究院数学研究所一年的回忆' (Remembering one year in the Institute of Mathematics of Academia Sinica), *Gannan shifan xueyuan xuebao (Ziran kexue)* 赣南师范学院学报

（自然科学）1989(S2): 1–3; reprinted in Chern Shiing-Shen (1989) *Chen Xingshen wenxuan* 陈省身文选, Beijing: Kexue chubanshe, v–x.

——(1989b) 'On the foundation of algebraic differential geometry', *Systems Science and Mathematical Sciences*, 2 (4): 289–312. Online. Available HTTP: <http://www.sysmath.com/xtkxysx_en/qikan/manage/wenzhang/jssc-89-2(4)-289.pdf> (accessed 16 June 2011).

——(1990a) 'On the chemical equilibrium problem and equations-solving', *Acta Mathematica Scientia*, 10 (4): 361–75.

——(1990b) 'On a projection theorem of quasivarieties in elimination theory', *Chinese Annals of Mathematics*, 11 (2): 220–26.

——(1992) 'On a finiteness theorem about optimization problems', *MM Research Preprints*, 8: 1–18.

——(1993) 'On the development of polynomial equations solving in China', in 吴文俊 Wu Wen-Tsun and 胡国定 Hu Guoding (eds) *Computer Mathematics: Proceedings of the Special Program at Nankai Institute of Mathematics Tianjin, China January–June 1991*, Singapore: World Scientific Publishing.

——(1996) *Wu Wenjun lun shuxue jixiehua*, 吴文俊论数学机械化 *(Wu Wen-Tsun on the Mechanization of Mathematics)*, Jinan: Shandong jiaoyu chubanshe.

——(1999) 'Zai "Qingzhu Shuxue tianyuan jijin chengli shi zhounian baogao hui" shang de jianghua 在"庆祝数学天元基金成立十周年报告会"上的啕话' (Speech at the Conference on the tenth anniversary of the Tian yuan Mathematical Foundation), in 胡作玄 Hu Zuoxuan and 石赫 Shi He (eds) *Wu Wenjun zhi lu* 吴文俊之路, Beijing: Kexue chubanshe.

——(2000) *Mathematics Mechanization. Mechanical Geometry Theorem-Proving, Mechanical Geometry Problem-Solving and Polynomial Equation-Solving*, Beijing / Dordrecht: Science Press / Kluwer Academic Publishers.

——(2002) 'Some reflections on mechanization of mental labour in the computer age', *MM Research Preprints*, 21: 1–5. Online. Available HTTP: <http://www.mmrc.iss.ac.cn/pub/mm21.pdf/wu.pdf> (accessed 9 June 2011).

——(2003) *Shuxue jixiehua*, 数学机械化 *(Mechanization of Mathematics)*, Beijing: Kexue chubanshe.

——(2004) 'Tansuo yu shijian – wo de kexue yanjiu licheng 探索与实践 – 我的科学研究历程' (Explorations and practice – the course of my scientific research), in 余翔林 Yu Xianglin and 邓勇 Deng Yong (eds) *Kexue de liliang (The Power of Science)* 科学的力量, Beijing: Kexue jishu chubanshe.

——(2005) 'Jisuanji shidai de naoli laodong jixiehua yu kexue jishu xiandaihua 计算机时代的脑力劳动机械化与科学技术现代化' (Mechanization of mental labour in the era of computers and the modernization of science), *Keji yu chuban* 科技与出版 *(Science, Technology and Publishing)*, 2005 (1): 29–31.

——(2006) 'Memory of my first research teacher: The great geometer Chern Shiing-shen', in P.A. Griffiths (ed.) *Inspired by S.S. Chern: A Memorial Volume in Honour of a Great Mathematician*, Singapore: World Scientific Publishing.

——(2008) 'Zhongyi ben xuyan 中译本序言 (Preface to the Chinese edition)', in B.L. van der Waerden (ed.) *Daishu jihe yinlun (Introduction to Algebraic Geometry)* 代书几何引论, 2nd edn, Beijing: Kexue chubanshe.

——(ed.) (1988c) *Xiandai shuxue xin jinzhan*, 现代数学新进展 *(New Advances in Modern Mathematics)*, Hefei: Anhui keji chubanshe.

——吴文俊 and Jiang Jiahe 江嘉禾 (1962) 'Essential equilibrium points of *n*-person non-cooperative games', *Scientia Sinica*, 11 (10): 1307–22.

Wu Wen-Tsun and Reeb, G. (1952) *Sur les espaces fibrés et les variétés feuilletés*, (On fibre spaces and foliations), *Actualités scientifiques et industrielles 1183 – Publications de l'Institut de mathématique de l'Université de Strasbourg*, Paris: Hermann & Cᵢᵉ.

Wu Wen-Tsun, Wang Dongming and Jin Xiaofan (1994) *Mechanical Theorem Proving in Geometries*, Vienna: Springer-Verlag.

Wu Wen-Tsun 吴文俊 and Wang Qiming 王启明 (1978) 'Theory of I*-functor in algebraic topology – I*-functor of a fiber space', *Scientia Sinica*, 21: 1–18.

Wu Xinmou 吴新谋 (1956) 'Canjia Bolan pianwei fangcheng huiyi (1955) ji 参加波兰偏微方程会议（1955）记' (Notes from the meeting on partial differential equations in Poland (1955)), *Shuxue jinzhan* 数学进展 *(Advances in Mathematics)*, 2 (2): 305–7.

Wu Yubin 吴裕宾 (1987) '*Jiu zhang suan shu* yu Zhongguo chuantong shuxue de fazhan 九章算术与中国传统数学的发展' (*Nine Chapters of Mathematical Techniques* and the Development of Traditional Chinese Mathematics), presented at *International Symposium on the 740th anniversary of Qin Jiushao's 'Mathematical Treatise in Nine Sections'*, Beijing. (unpublished). Quoted from NRI Offprint Collection, Mathematics XIV.

——吴裕宾 and Chen Xinghua 陈兴华 (1976) 'Wo guo zhongshiji jiechu de shuxuejia – Li Ye 我国中世纪杰出的数学家 —— 李冶' (An outstanding medieval mathematician of our country – Li Ye), *Shuxue de shijian yu renshi* 数学的实践与认识 *(Mathematics in Practice and Theory)*, 6 (4): 6–11.

Xi Zezong 席泽宗 (2000) 'Zhongguo kexue jishu shi xuehui 20 nian 中国科学技术史学会20年' (Review of the history of the Chinese Society for the History of Science and Technology in the last twenty years), *Zhongguo keji shiliao* 中国科技史料 *(Historical Materials of Chinese Science and Technology)*, 21 (4): 289–96.

Xi Ze-zong and Po Shu-jen [Xi Zezong 席泽宗 and Bo Shuren 薄树人] (1966) 'Ancient oriental records of novae and supernovae', *Science*, 154 (3749): 597–603.

Xi Zezong 席泽宗 and Guo Jinhai 郭金海 (2011) *Xi Zezong koushu zizhuan,* 席泽宗口述自传 *(An Oral Autobiography of Xi Zezong)*, Changsha: Hunan jiaoyu chubanshe.

Xinhua News Agency 新华社 (1984) 'Yuan Zhongguo kexueyuan Xinli yanjiusuo fu suozhang Fan Fengqi tongzhi yiti gaobie yishi zai Jing juxing 原中国科学院心理研究所副所长范凤岐同志遗体告别仪式在京举行' (Funeral ceremony of Fan Fengqi, former Deputy Director of the Institute of Psychology of CAS, in Beijing), *Renmin ribao* 人民日报 *(People's Daily)*, 23 July 1984: 3.

——(2001a) '2000 nian guojia zui gao kexue jishu jiang huojiang zhe Wu Wen-Tsun 2000年国家最高科学技术奖获奖者吴文俊' (Wu Wen-Tsun, winner of the Highest National Science and Technology Award for 2000), *Xinhua News*, 19 February 2001.

——(2001b) 'Rang chuangxin chengwei quan shehui gongtong de fengshang 让创新成为全社会共同的风尚' (Let innovation become a fashion for the whole society), *Xinhua News*, 19 February 2001b. Online. Available HTTP: <http://news.xinhuanet.com/china/htm/20010219/371420.htm> (accessed 4 September 2011).

Xiong Jincheng 熊金城 (2010) 'Huiyi shicong Wu Wenjun jiaoshou de rizi 回忆师从吴文俊教授的日子' (Remembering the days of study under professor Wu Wen-Tsun), in Jiang Boju 姜伯驹, *et al.* (eds) *Wu Wenjun yu Zhongguo shuxue* 吴文俊与中国数学 *(Wu Wen-Tsun and Chinese Mathematics)*, Singapore: Global Publishing.

Xu Chi 徐迟 (1978) 'Gedebahe caixiang 哥德巴赫猜想' (The Goldbach Conjecture), *Renmin wenxue* 人民文学 *(People's Literature)*, 1978 (1): 53–68.

Xu Lizhi 徐利治, Yuan Xiandong 袁向东 and Guo Jinhai 郭金海 (2009) *Xu Lizhi fangtan lu,* 徐利治访谈录 *(Interview with Xu Lizhi)*, Changsha: Hunan jiaoyu chubanshe.

Xu Zuzhe 徐祖哲 (2010), 'Xunfang quanqiu xunmi Min Naida xiansheng de shiji ' 巡访全球寻觅闵乃大先生事迹 (Looking for traces of Min Naida around the world). Online. Available HTTP: <http://xuzuzhe.blshe.com/post/6660/495359> (accessed 28 June 2011).

Yang Le 杨乐 and Li Zhong 李忠 (1996) *Zhongguo shuxue hui 60 nian,* 中国数学会60年 *(Sixty Years of the Chinese Mathematical Society)*, Changsha: Hunan jiaoyu chubanshe.

Yang Wenli 杨文利 and Zhang Meng 张蒙 (2007) 'Xin Zhongguo di yi ge keji fazhan guihua de zhiding, shishi ji lishi jingyan 新中国第一个科技发展规划的制定、实施及历史经验' (Establishment, implementation and the historic significance of the first development plan of science and technology in the New China), *Zhong Gong dang shi yanjiu* 中共党史研究 *(Studies in the History of the Communist Party of China)*, 2007 (6): 42–29.

Yang Xujie 杨虚杰 (2005) 'Wu Wenjun qujie beihou de meli gushi 吴文俊曲解背后的美丽故事' (A beautiful story behind the misinterpretation of Wu Wen-Tsun), *Chuangxin keji* 创新科技, 2005 (10): 32–3.

Yao Shuping (1989) 'Chinese intellectuals and science: A history of the Chinese Academy of Sciences (CAS)', *Science in Context*, 3 (2): 447–73.

Yau Shing-Tung, *et al.* (2011) 'Shiing-Shen Chern (1911–2004)', *Notices of the AMS*, 58 (9): 1226–49. Online. Available HTTP: <http://www.ams.org/notices/201109/rtx110901226p.pdf> (accessed 22 September 2011).

Yo Ging-Tzung 岳景中 [Yue Jingzhong] (1963) 'Secondary imbedding classes', *Scientia Sinica*, 12 (7): 1072.

Yu Yanlin 虞言林 (2010) 'Zhuhe Wu Xiansheng jiu shi huadan 祝贺吴先生九十华诞' (Congratulations to Mr Wu's 90th birthday), in Jiang Boju 姜伯驹, *et al.* (eds) *Wu Wenjun yu Zhongguo shuxue* 吴文俊与中国数学 *(Wu Wen-Tsun and Chinese Mathematics)*, Singapore: Global Publishing.

Yuan Zhendong 袁振东 (2008) '1978 nian Quan guo kexue da hui – Zhongguo dangdai keji shi shang de lichengbei 1978 年全国科学大会 – 中国当代科技史上的里程碑' (The National Science Conference in 1978 – A milestone in the history of science and technology of contemporary China), *Kexue wenhua pinglun* 科学文化评论 *(Science and Culture Review)*, 5 (2): 37–57.

Zarrow, P.G. (2006) *Creating Chinese Modernity: Knowledge and Everyday Life, 1900–1940*, New York: Peter Lang.

Zeng Zhaolun 曾昭抡, *et al.* (1957) 'Duiyu youguan woguo kexue tizhi wenti de ji dian yijian 对于有关我国科学体制问题的几点意见' (Some suggestions on China's science system), *Guangming ribao* 光明日报, 9 June 1957.

Zhang Baichun 张柏春 (2001) 'Dui Zhongguo xuezhe yanjiu keji shi de chubu sikao 对中国学者研究科技史的初步思考' (Preliminary reflections on the studies of the history of science and technology by Chinese scholars), *Ziran bianzhengfa tongxun* 自然辩证法通讯 *(Journal of the Dialectics of Nature)*, 23 (3): 88–94.

Zhang Dianzhou 张奠宙 (1999) *Zhongguo jin-xiandai shuxue de fazhan,* 中国近现代数学的发展 *(The Development of Mathematics in Modern China)*, Shijiazhuang: Hebei kexue jishu chubanshe.

Zhang Jingzhong 张景中 (2010) 'Ganxie he xuexi 感谢和学习' (Gratitude and emulation), in Jiang Boju 姜伯驹, *et al.* (eds) *Wu Wenjun yu Zhongguo shuxue* 吴

文俊与中国数学 *(Wu Wen-Tsun and Chinese Mathematics)*, Singapore: Global Publishing.

Zhang Meifang 章梅芳 (2003) ' "Pi Lin Pi Kong" yundong qijian Zhongguo ren de keji shi guan "批林批孔"运动期间中国人的科技史观' (Chinese conception of history of science and technology during the 'Anti-Lin Anti-Confucian' Campaign), *Kexue jishu yu bianzhengfa* 科学技术与辩证法 *(Science, Technology and Dialectics)*, 20 (3): 48–52.

Zhang Sucheng 张素诚 (1958) 'Jihexue yanjiu duixiang 几何学研究对象' (The objects of geometry), *Shuxue tongbao* 数学通报, September 1958: 396–7.

Zhang Xianfeng 张显峰 (2011) 'Wu Wen-Tsun: "Zuobie" shuxue de shuxuejia 吴文俊：“作别”数学的数学家' (Wu Wen-Tsun, a mathematician 'doing farewell' to mathematics), *Keji ribao* 科技日报, 17 August 2011. Online. Available HTTP: <http://www.stdaily.com/kjrb/content/2011-08/17/content_338132.htm> (accessed 4 September 2011).

Zhang Zhihui 张志辉, *et al.* (2008) 'Da zhi wei shi – Wu Wenjun xiansheng fangtan lu 大智为师 – 吴文俊先生访谈录' (An interview with Mr. Wu Wenjun), *Kexue wenhua pinglun* 科学文化评论 *(Science and Culture Review)*, 5 (5): 94–106.

Zheng Zhifu 郑之辅 and IMCAS (1964). *Ganbu jianding biao* 干部坚定表 (Cadre evaluation form), 20 March 1964, CAS Archives, file Z370-99/1.

Zhou Ping 周萍 (n.d.) 'Geming jia, shuxue jia Sun Keding 革命家，数学家孙克定' (The revolutionary and mathematician Sun Keding), *Jiangsu Sheng Wuxi Shi Di san gaoji zhongxue xiaoyou jianjie* 江苏省无锡市第三高级中学校友简介. Online. Available HTTP: <http://www.wxsgz.com/gk/showarticle.asp?articleid=1803> (accessed 6 October 2011).

Zhu Guangyuan 朱永远 and Li Xin 李新 (2005) 'Tan shen bo jing, da xiang wu xing – ji geming jia, shuxue jia Sun Keding 潭深波静 大象无形 – 记革命家、数学家、珠算家孙克定' (When a pool is deep, waves are tranquil; when a form is great, it has no shape – remembering the revolutionary, mathematician and abacus specialist Sun Keding), *Zhusuan yu zhuxinsuan* 珠算与珠心算 *(Abacus and Mental Arithmetic)*, 2005 (3): 44–8. Online. Available HTTP: <http://www.hljszx.com/page071008a.htm> (accessed 6 October 2010).

Zhu Qingshi 朱清时 and Jiang Yan 姜岩 (2004) *Dongfang kexue wenhua de fuxing*, 东方科学文化的复兴 *(The Rebirth of Oriental Science Culture)*, Beijing: Beijing kexue jishu chubanshe.

Zhu Rongji 朱镕基 (2001) 'Tuijin keji chuangxin, zaojiu jiechu rencai 推进科技创新造就杰出人才' (Promote innovation in science and technology, create excellent talents), *Xinhua News*, 20 February 2001. Online. Available HTTP: <http://news.xinhuanet.com/china/htm/20010220/372186.htm> (accessed 4 September 2011).

Zhu Shijie 朱世杰 [1303] (2007) *Si yuan yu jian jiaozheng,* 四元玉鉴校证 (Collated edition of the *Jade Mirror of the Four Unknowns*), edited by Li Zhaohua 李兆华, Beijing: Kexue chubanshe.

Index